THE SURVIVAL WORLD OF
BIRDS

First published in the UK 1992 by
Boxtree Ltd
36 Tavistock Street
London WC2E 7PB

1 3 5 7 9 10 8 6 4 2

© (text) John Gooders 1992

Design by Sarah Hall
Illustrations by Raymond Turvey

Typeset by Cambrian Typesetters, Frimley, Surrey
Origination by Fotographics, Hong Kong
Printed and bound in Hong Kong
by Dai Nippon Printing Company Ltd

A catalogue record for this book is available
from the British Library

ISBN 1 85283 112 X

THE SURVIVAL WORLD OF
BIRDS

JOHN GOODERS

ILLUSTRATIONS BY RAYMOND TURVEY

ANGLIA
Television Limited

B⬛XTREE

ACKNOWLEDGEMENTS

Any book, such as this one, that attempts to splash paint over a large canvas, must inevitably draw information from a wide range of sources. For this reason it is customary to include a bibliography. Yet to list all the various references that have been consulted while writing this book would be to miss a whole host of publications that have, over the years, been assimilated and remembered only during the writing process. It is easy to think of an "original" idea, only to discover the source at a later date. Similarly, what an individual knows about birds has been accumulated and made one's own. Therefore, to all of the authors of papers and books that I have read and forgotten, I offer my sincere, but unspecified, thanks.

Nearly twenty years ago I worked for *Survival* as a script writer under the watchful eyes of Aubrey (now Lord) Buxton and Colin Willock. There I learned the skill of communicating with a massive non-specialist audience. It is, I trust, a skill that I have never lost and never will. The skill is not, however, just to use layman's language, but to explain complex concepts in a way that can easily be understood. This is, after all, the communicator's business. So I thank *Survival* not only for choosing me to write this book, but also for engendering the skills required to produce the goods.

While I was at *Survival* I was fortunate enough to meet many of the photographers whose work graces these pages. I shall never forget Des and Jen Bartlett's *Flight of the Snow Goose* which I watched one snowy Christmas Day in Huddersfield. I travelled to Nepal with Dieter Plage on a reconnaissance that was to lead to an extended *Survival* project in that marvellous country and a really magnificent film on the tiger. Similarly, I was fortunate enough to visit Alan Root at his lakeside home on my first trip to Kenya and be shown in half an hour more than twenty birds that I have never seen before. For a short while it seemed that I was destined to spend some time marooned on the Bass Rock with Cindy Buxton, but then she went off to South Georgia and was marooned there during the Falklands War instead. These are the masters who produced the raw material which I, and others like me, sat and watched in the darkened cutting rooms at *Survival*. To them all I offer my thanks for an intensive course in the art of the possible and a glimpse of the impossible.

On a completely different, but no less valuable, level I have always been fortunate enough to be able to "talk" birds with a host of friends and acquaintances, many of whom are at the top of their profession. Again, like the missing bibliography, it is difficult to pick out what was gleaned from whom so I will simply thank them all for their help, in particular, Allan Keith (Chairman of the ABA), Stuart Keith (ABA's former Chairman), Peter Alden (formerly of Massachussetts Audubon Society) and the late James Fisher, Europe's first major bird popularist.

I must also thank my wife Robbie for the support and understanding which she has given me during the writing period, and Marion Waran who has now put more of my words on to the magic floppy disk than any previous secretary. I was fortunate to be able to dicuss the evolution of birds with an old friend, Cyril Walker, although all mistakes and contentious conclusions are entirely my own. In the final reckoning, the author is only a cog in the publishing industry and I would particularly like to thank Elaine Collins and Stephanie Walsh for leaving me alone during the gestation period and then bringing such expert and sympathetic hands to the editing process.

John Gooders, Winchelsea, Sussex

PREFACE

Many books have been written which catalog the details of each of the 157-odd bird families of the world. This book, however, takes a completely different tack.

Here, the focus is on zoogeography and the origins of the distinct avifaunas that exist on each of the world's great continents. It explains the differences and similarities; the way in which quite distinct groups of birds have evolved separately to fill similar ecological niches. It delves into the fascinating zones where different faunas collide and intermingle. It seeks to explain why some birds are confined to tiny areas, while others span the globe. Could it be, we shall ask, that some birds are as ancient as the continents themselves?

This comparison of the birds of the continents has proved an exciting task. Inevitably it has led to speculation in areas where there is no accepted wisdom and even to a closer examination of those areas where recognized explanations do exist.

Researching birds from the viewpoint of zoogeography has been very interesting and I hope that this new approach to the study of the birds of the world will be as enjoyable to read as it has been to write.

As this book goes to press, the ornithological world has been hit by the bombshell of DNA-DNA hybrid analysis, a technique that compares life forms on the basis of genetics rather than morphology. Just what effect this technique will have on avian systematics remains to be seen. The authors Sibley and Monroe do, for example, place the falcons in the same order as the penguins, along with groups as varied as the divers, flamingoes and vultures. Such an amalgam may take a bit of swallowing, but only time will tell whether the results are acceptable to the world's ornithologists. Meanwhile, the speculative nature of this book should, if nothing else, open the eyes of birders attempting to place their local birds in a world avifauna.

John Gooders, 1992

CONTENTS

INTRODUCTION

Birds are fragile creatures designed for flight. As a result their anatomy and physiology are totally geared to lightening their load at takeoff. Feathers, hollow honeycombed bones and a swift digestive system are all lightweight adaptions. It is therefore not surprising that compared with the dinosaurs with which they once shared the planet their fossil record is, to say the least, decidedly thin. Fossilized remains of the first known bird date from about 140 million years ago, but there then follows a gap of 15 million years before the next known bird. Ornithologists would love to know what happened during those missing 15 million years, but they are equally curious about the years before, during which, it is assumed, birds evolved from small reptiles.

The lack of a comprehensive fossil record is a decided handicap when attempting to see just how the birds of the world evolved. How is it, for example, that each of the southern continents (not counting, of course, Antarctica) has at least one large flightless bird? The rheas of South America, the African Ostrich, and the Emu and the Double-wattled Cassowary of Australia clearly share many features. But do they share a common ancestor, or have they simply evolved quite separately to fill a similar ecological niche? Certainly, there are plentiful examples of birds, remarkably similar both in structure and behavior, that have undoubtedly evolved quite separately. The hummingbirds of the Americas and the sunbirds of Afro–Asia are an excellent case in point. So too are the southern penguins and the northern auks. But these large flightless birds, the ratites, have been the subject of heated debate simply because the continents on which they live have been separated for longer than birds have been proved to exist on our planet. Thus, to postulate a common ancestor for the Ostrich, Emu and so on requires a total revision of our knowledge on the date when birds first evolved.

There is, of course, one further explanation. That is, that all of these large flightless birds – and to our existing list we should add the recently extinct

Despite its inability to fly, the Ostrich remains one of the world's fastest birds, capable of running at up to 40 mph. Such speeds are essential to a terrestrial bird that shares its savannah home with predatory lion, leopard and hyena.

elephant birds of Madagascar and the moas of New Zealand – must have evolved from a common ancestor that was able to fly.

If it seems strange to begin a book, which attempts to guide the reader through the variety and magnificence of the world's birds, with a somewhat obscure and puzzling problem rather than a straightforward statement of fact, I make no apologies. The world of birds is complex, but in that fact lies its fascination. This book is thus more than a catalog of wonders. Its aim is to compare groups of birds from the different continents, to see how they are similar and how they differ. And to seek an understanding, to pose questions and to open the mind to the dynamics that have shaped, and continue to shape, the most variable and exciting group of animals on earth.

Two distinct tactics have been evolved in the similar, but unrelated, penguins and auks. The Gentoo Penguin (above) of the southern oceans has dispensed with flight entirely, its wings becoming highly efficient flippers. The Tufted Puffin (left) "flippers" underwater too, but retains the power of flight with its short, rounded wings – a compromise solution between efficient feeding and escape from predators.

CHAPTER ONE

IN THE BEGINNING

The hard, horn-like crest of the Dwarf, or Bennett's, Cassowary is ideally suited to crashing through the dense undergrowth of its native New Guinea forests. Though kept as pets by local people, they are acknowledged as dangerous adversaries not to be trifled with.

Opposite: *Painted Storks at their nests at Bharatpur, northern India, where several thousand pairs breed in specially planted trees. Though a small family of only 17 species, the storks are found in every faunal region of the world.*

The limestones of Bavaria consist of exceptionally fine-grained material laid down during the Upper Jurassic period. They were formed as a sediment on the beds of shallow lakes and marshes covering the region at that time, and they remained untouched for some 140 million years. Their exploitation began only during the early part of the nineteenth century when it was discovered that these limestones made perfect plates for printing by the emerging process of lithography. Even so, each plate of limestone had to be free from imperfections, and a careful inspection process rejected any slabs that were uneven, cracked or contained impurities such as fossils. In no time at all Bavarian limestone acquired a reputation for producing outstanding fossils. Among the masses of plants, invertebrates and fish, all beautifully preserved in their fullest detail, there were also large numbers of reptiles including, in particular, the flying pterodactyls.

Then, in August 1861, quarry workers discovered the fossil remains of a feather and, a month later, an incomplete fossil skeleton was found. Named *Archaeopteryx lithographica* by Hermann von Mayer, these fossils were the first evidence of the existence of birds during the Jurassic period. Even today *Archaeopteryx* predates the remains of any other known bird by 15 million years. Several other *Archaeopteryx* have since been discovered, all of which are now preserved in European museums.

Momentous as the discovery undoubtedly was, its importance to us, in what is intended as a survey of the living birds of the world, is that it is the first bird that can be accurately dated. (Some scientists have claimed priority for other fossil specimens, such as Dr. Sankar Chatterjee's *Protoavis* which could be 75 million years older than *Archaeopteryx*, but these claims are still open to doubt.) *Archaeopteryx* was the size of a crow and lived about 140 million years ago. It may not have been the first bird, but it is certainly the earliest bird whose existence is beyond doubt. This, as we shall see, may be important in our survey of living birds.

The world 140 million years ago was, to understate the case, somewhat

One of only a handful of complete Archaeopteryx *fossils ever to have been found. Generally regarded as the first known bird, detailed examination shows several quite clear reptilian features, notably a toothed jaw, a long vertebrate tail and clawed fingers.*

different to the one we know today. It consisted of one, or perhaps two huge continents called Laurasia and Gondwanaland that, just a few million years before, had been joined together in the megacontinent of Pangaea. Gradually these two great landmasses broke up and drifted apart to form the continents as we know them today. One of the last "pieces" to slip into place was India which, having broken away from Madagascar and Africa leaving the debris of the Seychelles behind, drifted northeastwards until it crashed into Asia to form the Himalayas.

The idea of continental drift is widely accepted and there is reasonable agreement among experts as to an approximate time scale. The northern continent of Laurasia broke up less than 65 million years ago to form North America and Eurasia and it would be reasonable to search for similarities in both fauna and flora between the two. Gondwanaland broke up somewhat

The long toes of this Northern Jacana enable its weight to be spread over floating aquatic vegetation, as well as soft mud. As a result all jacanas have earned the nickname lilytrotters. Though closely related, these birds are found on each of the world's continents and have evolved into seven different species from a presumed common ancestor.

earlier and, although there must be a chance of finding similarities, it would seem unlikely that birds sharing a common ancestor could be found in South America, Africa, India and Australasia as the continents were already separating at the time that the first birds were evolving. Nevertheless, while the great northern continents of North America and Eurasia clearly do have a great deal in common and, indeed, are sometimes referred to jointly under the single zoogeographic name of the Holarctic, there are similarities in the southern continents, too. Each, for example, boasts at least one species of lilytrotter, or jacana, medium-sized marshland birds that have evolved distinctive long toes to facilitate running over floating aquatic vegetation. Though the species differ in plumage, ornithologists have placed them all in the single family Jacanidae, thus recognizing that they are closely related and share a common ancestor. If this is true, and there seems no reason to doubt it, then the original jacana probably lived in Asia and colonized the other continents by flight.

The same cannot be said of the ratites. This group consists of huge birds, including the largest in the world, that inhabit all of the southern continents. As they cannot fly they could not possibly have colonized these southern continents in their present form. Thus the rheas, Ostrich, Emu, cassowaries, elephant birds and moas must be descended from flying ancestors, or have evolved prior to the continents drifting apart. Although I have already mentioned the problems involved in exploring the evolution of the ratites, it is appropriate to examine them in greater detail here before attempting to follow the course of bird evolution in general.

Like the other ratite birds, these Emus of western New South Wales show plumage more akin to the fur of mammals than to the distinct feathering of flying birds. Once a species dispenses with the power of flight true feathers become redundant and are replaced by insulating, decorative, or cryptic plumage.

In a nutshell there are two basic theories put forward to explain the presence of these birds in such widely scattered regions of the southern hemisphere. The first is based on convergent evolution, the second on isolated evolution. The convergent evolution theory treats the ratites as a diverse and unrelated group of birds that, because of the similarities of their habitats, have evolved quite separately to resemble one another. They thus parallel other examples of convergent evolution such as the penguins and auks. In contrast, the isolated evolution theory regards them as closely related to a common ancestor that at some time in the distant past was able to fly and thus colonize the different continents where its descendants, adapted to local conditions, are found today.

Of the two theories the latter, postulating a common but flying ancestor, accounts for the considerable structural similarities between the various members of the group, especially noticeable in the structure of the palate, which is a feature that separates them from all other known birds. However, recent debate concerning the breakup of the southern super-continent Gondwanaland assumes a much shorter and more recent time scale than previously agreed. Separation of Africa and the Antarctic did not take place until 100–150 million years ago, while Africa remained attached to South America until some 80–90 million years ago and Australia drifted away from the Antarctic only 40–45 million years ago.

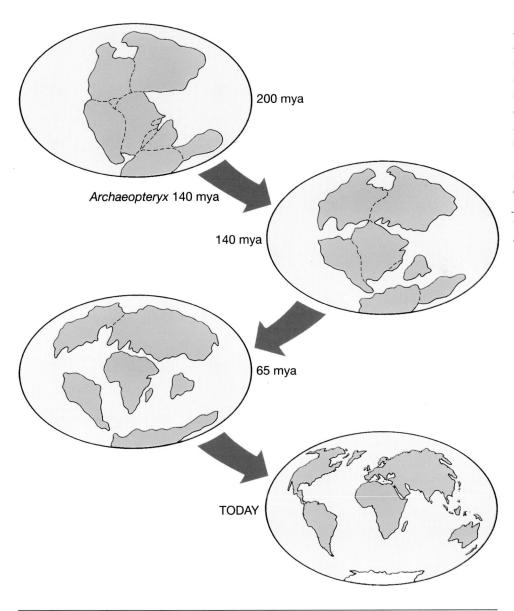

200 mya

Archaeopteryx 140 mya

140 mya

65 mya

TODAY

200 million years ago there was one megacontinent called Pangaea which gradually broke up to form the continents we know today; this process is known as continental drift. The evolution of Archaeopteryx and the timing of the break-up of Gondwanaland (the southern landmass) pose important questions about the distribution of the ratites, the large flightless birds, such as the Ostrich, which inhabit the southern continents.

American ornithologist Joel Cracraft from the University of Illinois has taken this revised time scale to posit a common nonflying ancestor for the ratites. This bird was once widespread in Gondwanaland and evolved into separate and isolated species as the continents drifted apart. Certainly, the revised time scale brings the breakup of Gondwanaland well into a time when birds were plentiful and therefore supports the single nonflying ancestor theory. The fact that such a theory enjoys a tidiness that is lacking in other theories does not mean that it is true. Indeed, paleontologists are skeptical not only of the theory, but also of its basis – the timing of the breakup of Gondwanaland.

I have spent time exploring the theory of ratite evolution largely because it picks out the most important premise of evolution in general. Species evolve from a common ancestor if, for one reason or another, they are geographically isolated for a period of time. Thus ornithologists make much of isolated islands such as the Galapagos and refer to them as "hot houses" of evolution. Certainly, this small group of islands off the coast of Ecuador is remarkable in many ways and was important to Charles Darwin in his preparation of *On the Origin of Species.*

The Galapagos are volcanic islands that emerged from the sea, some 600 miles from the nearest mainland, where no land existed before. Doubtless they were quickly colonized by seabirds and seaborne plant stems and seeds. No doubt, too, they were used by long-distance bird migrants as a resting place, and these also brought plant seeds with them. Yet fascinating as this colonization by mainland forms is, it was the arrival of a few rather nondescript mockingbirds and finches that captured the attention of Darwin. Originally the finches colonized different islands and, over a period, each adapted to the prevailing conditions by a process of natural selection. On an island colonized by thick-shelled seeding plants, the resident finches developed a deep, seed-cracking bill. In another location a thin, insect-picking bill was more advantageous. With geographical isolation on individual islands the original colonizing finch radiated into a variety of distinct species that became incapable of successfully inter-breeding. This was possible simply because the Galapagos were "new land" where a generalized feeder such as a finch had everything to itself and was able to adapt without competition from other existing bird specialists.

Undoubtedly the most extraordinary adaption was that of the Wood-pecker Finch, which developed a strong chisel-like bill and the remarkable skill of using a cactus needle to probe into insect holes. Mainland woodpeckers have developed special long, sticky tongues to search for food with. The Woodpecker Finch took a shortcut by changing its behavior.

Once a finch species had become sufficiently distinct from its relatives it became possible for it to spread to other islands where its specialized feeding and behavioral patterns prevented competition with an already resident species. If the two species were not sufficiently differentiated then one or the other would have died out.

The present-day population of Darwin's finches consists of 13 different species placed in four distinctive genera. They vary from the huge-billed Large Ground Finch to the delicately billed Warbler Finch. Most are dull-grey, sparrow-like birds, but some are pure black. Identification is based mainly on bill shape and bill size, for all still closely resemble their common ancestor in plumage pattern. The fact that some species bear a strong resemblance to an all-black (melanistic) phase of the widespread South American Bananaquit does not mean that this bird is the ancestor species. Convergent evolution may have produced such similarities and it is generally thought that the original ancestor species has become extinct.

That the process of evolution among Darwin's finches continues on the Galapagos can be gleaned from the distribution of the two cactus finches. For while the Cactus Finch is widely spread through the drier areas of all the major islands, the Large Cactus Finch is found only on four of the smaller outer islands and never on the same island as its more widespread relative.

Adaptive radiation of an initial colonizing species is seen to perfection in the Galapagos Islands of Ecuador. The chunky Woodpecker Finch (above) lacks the long prehensile tongue of woodpeckers and has evolved the unique habit of using a twig or cactus spine to aid its search for wood-boring insects. The so-called "Vampire" Finch (left), has followed another route to survival by pecking at nesting seabirds to feed on their blood and tissue.

The finches of the Galapagos are differentiated by their distribution, habits and bill structure. This Cactus Ground-finch feeds on the seeds of the opuntia cactus and has a medium-sized bill. Its closest relative, the Large Cactus Ground-finch, has a massive conical bill and feeds mainly on hard seeds gleaned from the ground.

While the evolution of the finches of the Galapagos is the most widely known example of adaptive radiation, the most spectacular example is to be found on another isolated island group, Hawaii. From a single ancestor the Hawaiian honeycreepers have developed into some 22 distinct species. Sadly, ten of these have already become extinct. One of the remaining species is the extraordinary Akiapolaau which has a woodpecker-like chiselling lower mandible and a thin decurved upper mandible used to extract insects like a woodpecker's tongue. Unlike the Galapagos the Hawaiian islands have been settled by man for a long time and the story of the destruction of natural vegetation and human exploitation makes depressing reading. Today most of the existing 12 species are confined to the forests that have survived high among the mountains where the introduced disease-carrying night mosquito cannot live. Some are nectar feeders like the Crested Honeycreeper which specializes on the flowers of the ohia lehua tree and is consequently rare. Others are insect eaters like the Akepa which has crossed mandibles to aid its probing among tightly-knit buds. The Palila has a thick seed-cracker of a bill and much prefers the hard seeds of the mamani tree, while the Laysan Finch is a more general feeder taking insects, seeds and even carrion and birds' eggs. As a result, while the

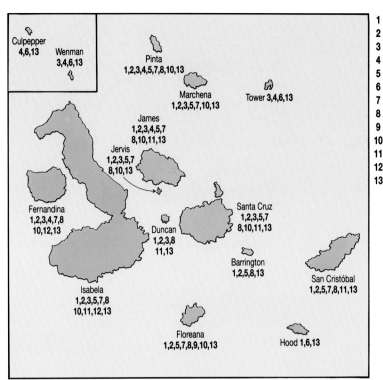

1 Small Ground Finch
2 Medium Ground Finch
3 Large Ground Finch
4 Sharp-beaked Ground Finch
5 Cactus Finch
6 Large Cactus Finch
7 Vegetarian Finch
8 Small Tree Finch
9 Medium Tree Finch
10 Large Tree Finch
11 Woodpecker Finch
12 Mangrove Finch
13 Warbler Finch

An examination of the distribution of Darwin's finches through the Galapagos Islands (above) shows how, from a common ancestor, the various species have evolved separately, but then been able to colonise elsewhere. Only by becoming different species can such similar birds co-exist on the same island. It is easy, for example, to see that the highly differentiated Warbler Finch exists on all the main islands, whereas the Medium Tree Finch is confined to only one.

The lower map highlights the distribution of the two species of cactus finch showing that the Large Cactus Finch cannot co-exist alongside the more widespread Cactus Finch. For different reasons the Mangrove Finch is found only where an appropriate wetland habitat exists.

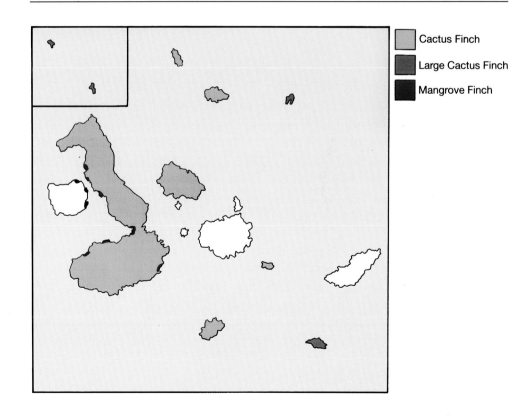

Cactus Finch

Large Cactus Finch

Mangrove Finch

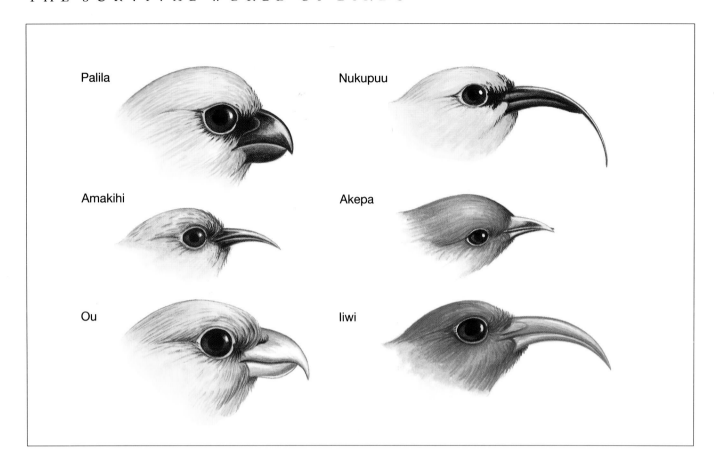

Palila

Nukupuu

Amakihi

Akepa

Ou

Iiwi

Perhaps the world's most spectacular example of adaptive radiation is found among the honeycreepers of the remote islands of Hawaii. On the left from top to bottom are Palila, Amakihi and Ou; on the right Nukupuu, Akepa and Iiwi. Such variation in bill shape shows how species can change to suit the particular environment in which they find themselves; all of these birds shared a common ancestor.

Palila is decidedly rare, the Laysan Finch, on the other hand, is relatively common.

It is in the nature of adaptive radiation that birds become differentiated by geographical isolation and subsequent specialization. The trouble is that specialists are unable to adapt to change, especially when change is as rapid as that inflicted on an area by man. Thus the Hawaiian honeycreepers are on the edge of a precipice with little chance of long-term survival.

It is easy to see that geographical isolation can be achieved on islands surrounded by miles and miles of hostile sea, but there are also many other ways in which isolation can occur. A bird that inhabits the open wastes of mountain tops may be isolated if there are no other mountain tops in view. The Kinabalu Warbler of Borneo, for example, finds itself isolated because its mountain home is surrounded by inhospitable forest. For this bird, forest is as inhospitable as the sea. Similarly, an isolated forest such as Sokoke forest on the coast of Kenya is so far from similar forests that no less than four distinct species – an owl, a pipit, a weaver and a swift – have evolved in glorious isolation and are not found anywhere else on earth. To these birds the surrounding plains are a sea of hostility.

One could continue with examples of birds that have evolved by virtue of being surrounded by areas that are alien to their needs, but the point is, I am sure, well taken – isolation is the key to the evolution of species. Perhaps not surprisingly, there is one further factor that is crucial to speciation – mobility, or rather the lack of it. Birds that breed in the polar and adjacent zones take advantage of the long days of summer and the flush of life that this provokes. Some may be able to eke out an existence during the winter, but most fly away to milder climes when their breeding grounds are in the grip of ice and snow. Such mobility promotes a strong "mixing" element, with individuals from widely scattered areas meeting and forming pairs. Such mixing has the effect of preventing geographical isolation by a few individuals and is thus a significant anti-speciation factor.

In contrast, birds that live in the tropics find the climate suitable to their needs throughout the year and have no need of migration. Consequently, they lack the "mixing" element and are more likely to enjoy some form of isolation and the propensity to speciate. It is not, therefore, surprising that a greater number of species can be found in the tropics than in temperate and polar regions. This is a somewhat simplistic view, however, for one of the prime features that affects the number of species found in any area is the diversity of habitats offered. Thus, even in the tropics, uniform areas are inevitably rather poor in the number of species to be found when compared with more variable areas.

In part, variety of habitats is dependent on underlying geology but this is heavily modified by temperature and rainfall. To illustrate this point, tropical rainforests are rich in species, but where they exist on the equator and grow over a wide altitudinal range – as, for example, among the Andes mountains of Ecuador – the variety of habitats and of bird species is quite extraordinary. In such an environment vegetation may lack a distinct flowering and seeding season, with trees of the same species being in flower throughout the year. As can readily be seen, such a routine allows for a high degree of specialization, to the extent that a bird species may specialize in feeding on a single species of flower. Similarly, wherever flowering trees are present in such a profusion of species, so too will be the birds that depend upon them. In such conditions even a comparatively small change in altitude will be sufficient to produce new species of both trees and birds – for example, among the rich cloud forests of the Pacific slope of Ecuador, where virtually every turn in the zigzag mountain road produces a new species of hummingbird. There are, of course, many more subtle influences at work in determining the richness or paucity of species found in any particular area, country or region, but latitude, rainfall and altitude are undoubtedly the most important.

The breakup of Laurasia and Gondwanaland and the subsequent drifting of the continents into their present positions closely corresponds with the division of the world into zoogeographic, or natural, regions. This is not really surprising as the evidence of the one is an important factor in the elucidation of the other. Present-day zoogeographers recognize six natural regions – the Palearctic, Nearctic, Ethiopian, Oriental, Neotropical and Australian regions. Antarctica is sometimes treated separately but, more often than not, lumped together with the Australian region. Some of these regions are clearly demarcated and delineated, others are divided more

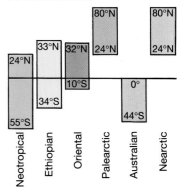

LATITUDE RANGE

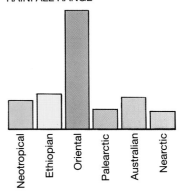

RAINFALL RANGE

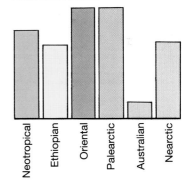

ALTITUDE RANGE

Basic geographical factors such as latitude, rainfall and altitude affect the richness or poverty of the world's zoogeographical regions. The latitude and altitudinal ranges of the Neotropical region are the main reason for its dominant position as the richest bird region of the world.

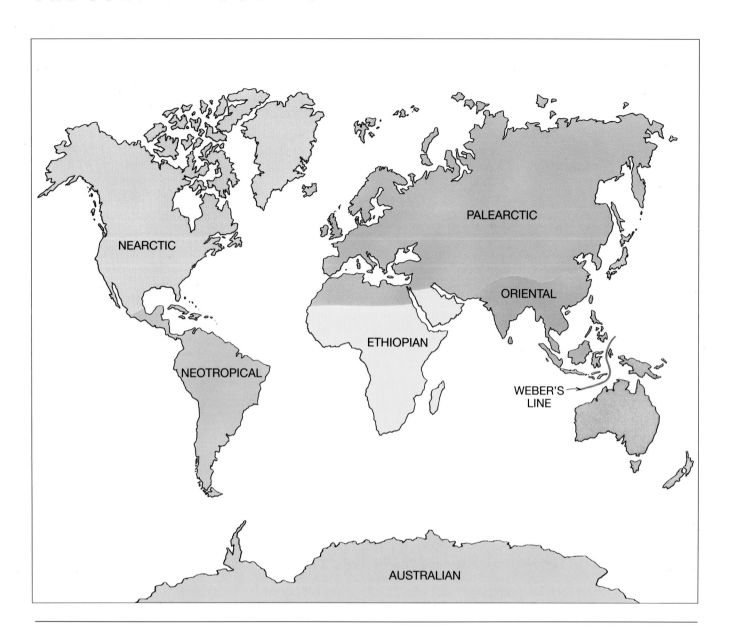

The six zoogeographical regions of the world, with Madagascar included in the Ethiopian region and Antarctica in the Australian region.

subtly. The Nearctic region, for example, consists of North America, and the Neotropical region consists mainly of South America. Where these two regions meet in Central America a detailed examination of fauna and flora is required to draw an exact line. The most classic division is "Wallace's Line", drawn by Alfred Russel Wallace, that divides the Oriental region from the Australian. It runs between Borneo and the Celebes and between Bali and Lombok. While these latter two islands are barely 20 miles apart they show the most extraordinary differences in terms of flora and fauna.

As the zoogeographical regions correspond closely to the continents themselves it is clear that their faunas and floras have evolved in isolation, the same isolation that, on a smaller scale, is responsible for the evolution of an individual species. Most of the zoogeographical regions are surrounded by inhospitable seas. Those that are not are separated by other equally

A Western Flycatcher at its tree-
hole nest site in Colorado. A
glance at a Neotropical field guide
is sufficient to show the
extraordinary species radiation of
the Empidonax flycatchers, of
which ten have colonized the
Nearctic region. All share the
basic plumage pattern that makes
them a birdwatcher's nightmare to
identify.

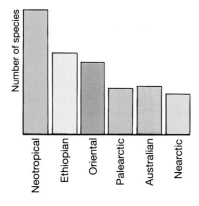

Number of species

Neotropical Ethiopian Oriental Palearctic Australian Nearctic

The Neotropical region has the greatest number of breeding species (2,500), followed by the Ethiopian region (1,600), the Oriental (1,400), the Palearctic and Australian (900 each) and lastly the Nearctic (750).

formidable barriers. The Oriental region is separated from the Palearctic by the Himalayas, the world's highest mountain chain. The Ethiopian region from the Palearctic by the Sahara, the world's largest desert.

Such isolation does not, of course, prevent birds from moving from one continent or region to another, for these are the most mobile of all the world's animals. What it does do is to provide the opportunity for species to evolve, even though other, perhaps related, species may use the area during part of the year. We can reasonably expect, therefore, that many species or species groups (such as genera, families or even orders) will be confined, or largely confined, to a particular region where they have evolved.

While isolation may be created by deserts and mountains, it is the sea that has proven the most effective at family level. Thus it is the Australian region that boasts the largest number of endemic bird families. These include the megapodes, magpie-larks, birds of paradise, bowerbirds, lyrebirds, scrub-birds and honeyeaters which, with the kiwis, wattlebirds and wrens of New Zealand, make a total of ten complete families that occur nowhere else on earth. While other regions may have a near monopoly of species of a particular family, totally endemic families are comparatively rare. The land link between North and South America, between the Nearctic and Neotropical regions is only a million or so years old, yet this period has been quite sufficient to allow a considerable interchange of species of different families. For example, typical Neotropical families such as the ovenbirds and tyrant flycatchers have spread northwards into the Nearctic while tanagers and pipits have spread southwards.

In the same way, although the Nearctic is separated from the Palearctic to the east by the huge North Atlantic, in the west it is only the narrow Bering Straits that offer a barrier. Even here there was, in comparatively recent times, a land bridge between the two regions. As a result, the Nearctic has few, if any, endemic families even though several families clearly have their origins there. The land link between the Palearctic in the north and the Oriental region to the south has enabled species to move between one and the other so that, save for the accentors which are confined to the Palearctic, neither region can boast endemic families.

Despite the high level of endemism in family terms to be found in the Australian region, the supremacy of the Antipodes does not continue at species level. In the leading position in respect of the number of breeding species to be found comes the Neotropical region with 2,500 species. Next comes the Ethiopian region with 1,600 species, followed by the Oriental with 1,400, the Palearctic and Australian with some 900 species each and finally by the Nearctic region with only 750 species. In the light of these figures it is perhaps not surprising that so many North American birders cast their eyes southwards towards the bird riches of South America where there are over three times as many different species to be seen as at home.

Though the zoogeographical regions form the basis of this survey of the world's birds it is an interesting aside to compare the avifaunas of different political areas or countries. At the top of the table comes Colombia with over 1,600 breeding species. Sadly, habitat degradation is so extensive in

Colombia and distances between good bird habitats so daunting that most birders prefer to visit neighboring Venezuela, where the national list is only about 1,300 species. Most European countries have lists of between 300 and 400 species, though the British Isles boasts over 500, mainly because of its geographical position at the edge of a continent where rarities are more likely to occur. The United States has about 670 breeding species, with Texas alone pushing towards the 600 mark. The smallest lists are found on isolated islands and among remote archipelagos, in desert countries and in the polar regions. It is, however, worthy of mention that although both the Arctic and the Antarctic are low on species, the numbers of birds found in these areas are among the highest anywhere. Indeed, the title of the world's most abundant bird may lie between Wilson's Petrel of the Antarctic and the Little Auk, or Dovekie, of the Arctic.

Little Auks, or Dovekies, gather in enormous numbers at their arctic breeding grounds. Though they migrate southwards to more temperate climes in winter, they remain well out to sea and are only rarely viewed from land, usually after strong autumn gales drive them to shore. Occasionally, severe gales may cause a "wreck" of birds inland.

CHAPTER TWO

BIRDS OF THE WORLD

A Northern Gannet with a billful of buttercups to present to its mate – not as a posy, but as mundane nesting material. The Gannet was one of the first of the world's birds to be accurately censussed and our knowledge of its population dynamics now covers a period of over 50 years.

Opposite: *Camouflaged to perfection, a female Red Grouse incubates her eggs on a Scottish moor. This bird is the quarry for what is probably the world's most expensive shooting.*

There are about 8,600 distinct species of birds to be found living in the world today. They are divided among about 2,100 genera and grouped into 157–159 different families. The vagueness of these totals is not a result of the inability of ornithologists to count, but of the nature of the subject matter itself. Birds, along with every other form of life on our planet, are evolving continuously. Some, quite naturally, are becoming extinct just as others are developing. The process is dynamic, yet the human sense of order attempts to achieve an impossible stability. We love making lists and many have attempted to list all of the world's birds. Each list differs from the others, not in its lack of knowledge, but in what actually counts as a species. An example may help to clarify this point.

The British Isles has for long been noted as the home of the endemic Red Grouse, a bird found nowhere else on earth. Ornithologists were agreed that this bird was very closely related to the Willow Grouse (Willow Ptarmigan in North America) but, lacking white wings and a special all-white winter plumage, the Red Grouse was regarded as a full species in its own right. More recently, plumage characteristics have given way to physical structure as a prime means of differentiating species and, even more recently, the high technology of genetics has come into play. As a result, the Red Grouse is now regarded as no more than a well-marked subspecies of the Willow Grouse and Britain has lost its only endemic bird.

Yet hardly had the one species fallen than these same islands were to be granted a replacement. The Crossbills that periodically invade Britain from mainland Europe and which sometimes, and increasingly, stay on to breed were, it was decided, quite different from the birds that breed and are resident in Scotland. The latter have heavier bills, more suited to dealing with the larger and tougher cones of the native Scots pine, than their Continental relatives, and ornithologists came to the conclusion that they were sufficiently differentiated to form a distinct species. Thus, ornithological opinion about two species that are members of one of the world's

Avian classification is based on a pyrimidal structure with individual species (at the bottom) grouped into genera, genera grouped into families, families into orders and orders into the single class Aves. The approximate number of these different categories is shown on the right. On the left we see that the Herring Gull Larus argentatus *is placed in the gene* Larus, *which in turn is placed in the family* Laridae, *then in the order* Charadiiformes.

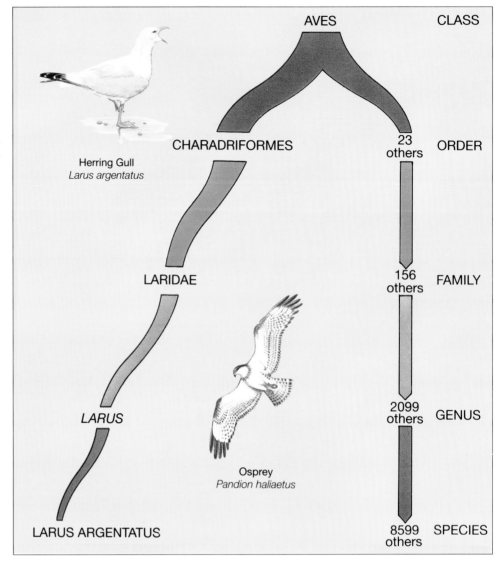

AVES — CLASS

CHARADRIFORMES — 23 others — ORDER

Herring Gull
Larus argentatus

LARIDAE — 156 others — FAMILY

LARUS — 2099 others — GENUS

Osprey
Pandion haliaetus

LARUS ARGENTATUS — 8599 others — SPECIES

best known avifaunas changed in a matter of a few brief years. Is it any wonder that opinion changes even more rapidly in less-well-understood regions? Ornithologists are, then, vague about numbers because the birds themselves are changing; but equally, and perhaps even more significantly, because ornithological opinion is itself subject to change.

There is a special group of zoologists and botanists called systematists. It is the job of the systematist to examine species and subspecies of animals and plants, seek out their relationships and attempt to bring a sense of order to nature that we all agree is impossible. The fact that it is impossible in no way invalidates the efforts made. For while there can never be a "correct" order, systematists can help us to understand better the relationships between different species of living things.

Systematists can be divided into two broad groups – "splitters" and "lumpers". Splitters are inclined to separate different forms into distinct species, while lumpers tend to lump together into a single species. Thus, when a splitter produces a list of the birds of the world the result may be up to 9,000 different species, whereas a lumper may produce a list of only 8,300. Yet, remarkably, most systematists arrive at a figure somewhere between 8,500 and 8,600 species. However, there have recently been radical developments in bird classification. The work of Charles Sibley, Burt Monroe and Jon Alquist, using the results of DNA–DNA hybridization studies to measure the genetic similarity between species, looks set to challenge all previous studies. Their revolutionary work may, at last, bring order to what has often been an ornithologist's nightmare.

Systematics has its origins in the mid-eighteenth century with the work of the remarkable Carl von Linné who Latinized his own name to Carolus Linnaeus. The third edition of his *Systema Naturae* in 1858 is accepted as the beginning of the Latin binomial system on which all subsequent work on nomenclature is based. Known as the Linnaean or scientific system, each species is given a two-part (thus binomial) Latinized name. The first part is the generic name, the second the specific name. Thus the scientific name of the Red Grouse was formerly *Lagopus scoticus*, while the Willow Grouse was (and still remains) *Lagopus lagopus*. Today there is only one species *Lagopus lagopus*. In this case the generic name *Lagopus* is the same as the specific name *lagopus*. By convention all scientific names are printed in italics and the generic name always starts with a capital letter.

As the Red Grouse is such a well-marked form of the Willow Grouse it has been accorded subspecific status via the system of Latin trinomials. It is named *Lagopus lagopus scoticus*, which differentiates it from the Willow Grouse *Lagopus lagopus lagopus*. For sheer convenience, and again by convention, consecutive mentions of scientific names in printed form are abbreviated. In this particular context the Willow Grouse becomes *L. l. lagopus* and the Red Grouse *L. l. scoticus*. In practice, however, subspecies with their cumbersome trinomials are seldom discussed and the Latin binomial *L. lagopus* is used to refer to the species.

In all, the 8,500–8,600 species of birds are grouped into approximately 2,100 genera. Each group is referred to as a genus. Some genera are huge, like the genus *Zosterops*, the white-eyes, with almost 60 recognized members. Others consist of only a single species like the genus *Pterocnemia*, the Lesser Rhea. While the former are referred to as polytypic, the latter forms a monotypic genus. Similarly, rather like a pyramid, genera are gouped into families of which there are some 157–159 depending on the authority or author concerned. In the case of the white-eyes no less than eleven different genera are grouped together to form the family Zosteropidae, whereas the rheas, the Rheidae, consist of only two genera.

Whereas the world's most primitive birds, the earliest to evolve, are grouped into genera and families with, like the rheas, comparatively few members, the more recently evolved birds, like the white-eyes, consist of large genera and large families. There are several bird families that are monotypic, that is, consisting of only a single species. Good examples are the Struthionidae or Ostriches; Dromaiidae or Emus; Balaenicipitidae or

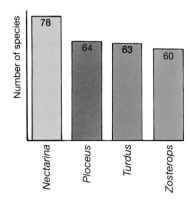

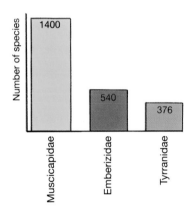

The bar chart above shows variations in the number of species among some of the larger genera of birds. The chart below shows the much larger numbers that are grouped into families. The huge family Muscipapidae contains no less than 16 per cent of the total species of birds in the world.

Shoebills; Scopidae or Hammerheads; Sagittariidae or Secretary Birds; Pandionidae or Ospreys; Opisthocomidae or Hoatzins; and the Pedionomidae or Plains Wanderers. Such examples of single species families are much rarer among the smaller perching birds but include the notable Palm Chat, the sole member of the family Dulidae.

While one cannot have a smaller family than one member, at the other end of the scale the number of species shows enormous variation. The ovenbirds of the family Furnariidae number over 260 species and the Emberizidae 540 species. Moreover, the family Muscicapidae consists of some 1,400 different species; the Old World flycatchers and warblers alone of this family represent no less than 16 percent of the world's birds.

It is remarkable that, despite all of the time and effort that has been afforded to the scientific names of birds, the widely used English names are still a complete jungle. Some species have two or more English names, like the Swallow and Barn Swallow or the Long-tailed Duck and Oldsquaw. Other species may have only a single English name, but one that is equally applied to a completely different species in another part of the world. I have long argued the case for an internationally accepted list of English names based on common usage with change or qualifiers where necessary. Others have wished to change English names in line with scientific names, thus reducing the American Robin, for example, to Red-breasted Thrush or some such. Still others see no need for change, allowing the development of a language, in this case English, to follow a natural evolutionary course with all the enriching and confusing benefits that ensue. While this latter course has much to recommend it, the premise that in the scientific name system we have a perfectly good, internationally understood set of names misses an important point, namely, that most people who are interested in birds are not professional ornithologists. Nevertheless, I would advise anyone with only a passing interest in the subject to become familiar with the scientific names of every bird they encounter, if only to understand a little more about its relationships with other species.

A point worth noting, for example, is that members of the same family are regarded as having descended from a common ancestor. It is not surprising, therefore, to find that all of the world's jacamars are found only in South and Central America, as indeed are all of the motmots. The world's five species of todies not only belong to the same family the Todidae, but also the same genus *Todus* and are restricted to the Caribbean. However, an equally small family, the flamingoes, is among the most widespread of all bird groups. Two species, the Andean and James, are found only among the high mountain lakes of the South American Andes and the Lesser Flamingo is restricted to the lakes of East Africa and western India, but the Greater Flamingo ranges across no less than five continents. Whatever its origins, probably subtropical, this remarkable bird is clearly a great wanderer for despite its extraordinary range only two subspecies are generally recognized.

An even more remarkable family is the monotypic Pandionidae. This family's single species, the Osprey, can be found in every corner of the

world and on every continent. It is as much at home on the lakes of North America as it is among those of Europe. It has found its way from Asia to the Philippines, to Australia and through the Pacific. Were it not for direct persecution by man and the indirect effects of worldwide overfishing, this magnificent and spectacular bird would be common virtually everywhere that fish can be found.

Another equally widespread bird is the Peregrine Falcon, which is more terrestially based than the Osprey, though neither are noted swimmers. Its strong flying abilities have enabled it to colonize all the continents and virtually the whole ice-free land surface of the earth. Likewise, its relatives of the genus *Falco* have also successfully colonized huge areas of the globe, though some have become resident and developed into species in highly restricted areas. The rare and much endangered Mauritius Kestrel is perhaps the best-known example, but there are also endemic kestrels in the Seychelles, Madagascar (two), and New Zealand.

Of all the bird families none, I would venture, has proved so susceptible to speciation as the Rallidae, the rails, crakes, coots and their allies. Though consisting of some 125 different species, many are closely related and placed

Although one of the most widespread of all the world's birds, the Osprey, seen here returning to its nest in Florida, has suffered a severe decline during the present century. At first this was due to direct persecution, but today pollution poisoning and decline in fish stocks are mainly to blame.

Above: *Lesser Flamingoes, like these flying at Etosha Pan, Namibia, are found only in the Ethiopian and Oriental regions. They are most abundant on the Rift Valley soda lakes of Kenya, where more than a million pairs breed.*

Right: *One of the rarest of all the world's birds, the Mauritius Kestrel survives only by the intensive efforts of a few dedicated conservationists, backed by European and North American fundraising. Once widespread, its local name of "chicken-hunter" played a significant part in its decline due to persecution.*

within the same genus. There are, for example, nine distinct coots, seven of which are found in the Americas while five of these are restricted to South America. Yet one species, the Common Coot, has managed to spread to every other landmass in the world. Its success has been remarkable.

Also very successful have been the members of the genus *Rallus*, the typical rails. The Water Rail is widespread across Eurasia, as is the Virginia Rail in the Americas. Indeed, despite their apparently poor flying abilities, rails of this genus have managed to colonize some of the most remote corners of the world. A catalog of living island rails makes for surprising reading, especially when it is borne in mind that many island forms have become extinct during the past 200 years. But the important point is that all these birds belong to the same genus and therefore shared a common ancestor not that long ago, in geological terms at least.

Platen's Rail is confined to the Celebes, while Wallace's Rail is found only on Halmahera Island and the New Britain Rail only on the island of the same name. The New Caledonian Wood Rail may be extinct on New Caledonia and, therefore, in the world, and the Lord Howe Wood Rail is restricted to that island. The Barred-wing Rail is confined to a group of islands around Guadalcanal, the Guam Rail is found only on the island of Guam and the Madagascar Rail is restricted to the eastern part of that large island. So, of 21 members of the genus no less than eight are restricted to islands or island groups. Their close relationship is incentive for us to seek an explanation of their distribution and an understanding of how species can evolve in isolation over such a short period of time.

In the same way as we find it interesting to compare closely related birds that have such a worldwide, or at least widespread, distribution, it is just as enlightening to check the "mirror image". By this, I mean to note the differences between the zoogeographical regions or continents, and to examine, and perhaps explain, why some species, genera and families are confined by their origins. Why, for example, are hummingbirds confined to the Americas? Why are they replaced in the Old World by the sunbirds? Why, of nearly 220 species of ovenbirds, have only 15 managed to colonize as far north as Central America and only two of these are found exclusively there? Why are the mousebirds found only in Africa south of the Sahara? And why are penguins not found in the Arctic as well as the Antarctic? How have several species of albatross managed to cross the doldrums and the equator to colonize the North Pacific, but none, save the odd individual, have done so in the Atlantic? Why are all of the world's 60-odd species of wren confined to the New World, save for one single species that has been able to spread successfully right across Asia to Europe and even to some of the most remote of North Atlantic islands? Why are there no dippers in Africa, Australia and New Zealand? The answers to these and a great many other questions will be given later. For now we can only wonder at the extraordinary variety of the birds of the world.

Birds differ not only in distribution, but also in the habitats in which they live, in their size and structure, in their coloration, in their behavior and nesting routines, in their seasonal patterns – in fact, in almost every

The Common Coot has managed to colonize virtually every significant landmass in the world, while seven out of eight of its closest relatives are confined to South America. Just what is responsible for the Common Coot's success is not known, though having managed to break away from the Neotropical region it must have found itself in a coot-free-zone totally lacking in competition.

conceivable way. There is, for example, one bird that never ever comes to land. The Emperor Penguin spends its life at sea and breeds on the sea when it is frozen in the depths of the Antarctic winter. The female lays a single egg that is then passed to the male, who rests it on his feet and covers it with a specially extendable flap of skin. While his mate returns to the sea to feed and recover from forming and laying the egg, the male incubates the egg through the worst weather this planet can offer. He survives two months without food and still has sufficient body reserves to feed the chick when it hatches. Then the female returns fully fit and cares for the growing youngster by producing an oily substance from her crop, while the male in turn feeds at sea. This routine is, perhaps, the most exceptional adaption to exceptional circumstances in the world of birds and demonstrates to perfection that birds have not only evolved physically to suit their environment, but also behaviorally.

If the Emperor Penguin never comes to land, the Ostrich never leaves it. Like the other ratite birds discussed earlier, the Ostrich is incapable of flight and has evolved a set of life tactics to survive among the plains of Africa. It is, for example, exceedingly large, indeed the world's largest living bird. A male may stand up to eight feet tall and weigh up to 350 pounds. It can run at over 40 miles per hour and is quite capable of breaking every human running record up to 1,500 metres and probably beyond. Despite its

inability to fly it is classed among the fastest of all birds. Adult Ostriches are consequently far from being easy prey for the great predators with which they share a home. They are, however, less well protected during the breeding cycle, and, as a result, the Ostrich has evolved a special reproductive tactic.

The male Ostrich uses his short black and white wings as flags to signal his presence over the open plains. This peculiar semaphore-like system proclaims his territory to other males and attracts females. With a harem of two to five females the male encourages them to lay all their eggs in the same nest. In this fashion, a clutch of 15–50 eggs may be produced that the male, helped by one or two females, then incubates. This can be contrasted with the single-egg tactic of the Emperor Penguin on its secure, but hostile,

Rails, like this Water Rail, are inhabitants of dense marshy vegetation and are generally skulking and difficult to see. Their long legs and toes are perfectly suited to their habitat and their thin, laterally compressed bodies pass easily through vegetation.

Even among large flightless birds, breeding tactics vary enormously. The Emperor Penguin (right) lays a single large egg in the security of the Antarctic winter, where predators are virtually non-existent. The even larger Ostrich (opposite below), in contrast, gathers a harem of females to produce a clutch of twenty or more eggs, simply because egg and chick predators are so numerous. Even the mighty lion (opposite above) is not below taking a fresh Ostrich egg for breakfast.

Antarctic ice floe. With each Ostrich egg being roughly equivalent in volume to 20 hen's eggs, the food value of such a clutch is all too obvious and many are lost to jackals and other predators. Over the past 30 years or so Egyptian Vultures have learned how to break Ostrich eggs by cracking them with stones. Having discovered an Ostrich egg, the vulture picks up adjacent stones with its bill and forcefully throws them to the ground. Usually it misses, but persistency eventually pays off, the egg is broken, and the contents make a splendid meal even for a relatively large bird like a vulture. Though the vultures are far from expert, this trick is one of the very few examples of tool-using among animals.

Of course, by the sheer law of averages sufficient eggs do survive to hatch, but then the quick-running chicks become immediate prey to a whole range of predators. Only by vigorous defense on the part of the adults and through their natural camouflage and ability to "freeze", do enough youngsters survive to keep the species going.

From the largest to the smallest, from the Ostrich to the tiny Bee Hummingbird, which at two and a half inches and weighing less than a tenth of an ounce is the world's smallest bird – in fact a great many insects are larger. This little waif beats its wings up to 80 times a second to produce a high-pitched, bee-like hum and can fly both forwards and backwards with apparent ease. Its metabolic rate is, however, highly expensive on fuel and virtually every moment of daylight is spent sipping nectar in order to keep itself alive.

Another group of birds that seems to spend most of its time feeding are the swifts and here too there are some quite outstanding adaptions. Unlike the Emperor Penguin and Ostrich, these are the most aerial of all birds and there is considerable evidence that, except for nesting, many species may spend virtually their whole year airborne. Food consists entirely of insects, a sort of aerial plankton that is gathered up in their huge mouths. Nesting materials are made up of feathers, straw and windblown debris that is then glued together with saliva.

Indeed, some of the smaller swiftlets construct their nests entirely of dried-out saliva, a fact that has led to their protection and conservation in the caves they inhabit in southeast Asia. For when the swiftlets have finished building, the nests are collected to form the basis of bird's-nest soup, an epicurean delight that is highly prized in the Orient. Recent research, however, has shown that the breeding success of birds forced to construct a second nest is lower than those that nest unmolested. Also, disturbance of their nesting caves can seriously reduce their productivity in rearing young, and ornithologists have therefore suggested a nondisturbance policy during the middle of the breeding cycle to ensure that the colonies remain both healthy and fruitful.

A halfway house between saliva and a proper, albeit aerially gathered, nest is that constructed by the Palm Swift. These slimline birds gather feathers and aerial debris, but then glue the mass to the underside of a palm frond. The single egg is then glued to the nest and the resulting chick clings to the nest as soon as it hatches. Like the other swifts it mates in the air, itself no mean achievement, and may well sleep in a series of aerial catnaps.

Although the swifts are undoubtedly the most aerial of birds, they are not the world's greatest travelers. That distinction rests with the remarkable Arctic Tern, which nests in the Arctic and winters in the Antarctic where it spends the southern summer feeding among the breaking pack ice – and, incidentally, perhaps circling the world. During migration alone the Arctic Tern covers more than 22,000 miles, or about double the annual mileage of the average motorist. Add in diversions, feeding forays and breeding season mileage and it seems likely that some birds must fly about 80,000 miles a year.

Despite being so far ranging, these terns still have quite distinct summer and winter distributions. They breed around the North Pole from the Bering Straits right across northern Canada northwards as far as there is ice-free land. They nest in Iceland, northern Britain and from Scandinavia across the full breadth of Russia and Siberia back to the Bering Sea. Yet on

migration they move southwards down the eastern side of the Atlantic and the eastern side of the Pacific, presumably benefitting from the fish-rich seas that exist along these coastlines. In winter, as we have noted, they are widespread around the Antarctic.

Despite such a huge circumpolar breeding distribution, the Arctic Tern boasts no subspecies, a clear indication that there is considerable "mixing" among the various populations. There is, however, a separate species, the Antarctic Tern, clearly closely related to the Arctic Tern, that inhabits some of the remote islands of the Antarctic region as well as parts of the continent itself. Some authorities regard this southern bird as a derivative of the northern bird and lump the two together to form a superspecies. Nevertheless, while the two spend much of the southern summer feeding alongside one another, they are separated by the length of the earth during the northern summer. That two closely related birds can live alongside each other in this way in the south raises the obvious question as to why they cannot do so in the north. In fact they do, but in this case the Antarctic Tern is replaced by the Common Tern, which while often breeding alongside the Arctic Tern does not migrate anywhere near so far south.

The mobility of the Arctic Tern can be contrasted with the highly residential behavior of birds such as the Red Grouse. Though it may live for several years, the average Red Grouse seldom moves more than a mile or so

Arctic Tern, the world's greatest traveler, covers more miles on migration alone than the average sales representative clocks up in a year. Yet by flying from one end of the earth to the other it enjoys longer hours of daylight than any other animal.

The Painted Stork of the Orient is a migrant within a single region. It breeds towards the end of the monsoon in northern India and then heads southward during the winter to milder climes. There it joined other Painted Storks that do not migrate at all. Such a regime is referred to as partial migration.

from where it was hatched. Individuals are spread evenly through the heather-clad moorland by virtue of their highly territorial behavior. Birds that cannot find and defend a territory are forced into unsuitable habitat and eventually die. There is thus a definite limit to the number of Red Grouse that a particular moor can accommodate, a situation that provides a splendid argument for those seeking to justify the sport of grouse shooting. It is, the argument goes, only the surplus birds that are shot each autumn, birds that would otherwise be driven from the moor to die a lingering death from starvation. All that man can do is to ensure that the moor is in good health to offer as much living space as possible for the Red Grouse. To this end the grouse moors are systematically burned on a rotation basis to provide the plentiful young heather shoots on which the birds depend.

Short, rounded wings have not prevented the Common or Winter Wren from becoming a long-distance migrant in its native North America. In western Europe it is largely resident and has evolved several distinct land subspecies as a result.

Because it is such a stay-at-home the Red Grouse has no need of prolonged flight, but it does still need to escape predators and has not lost the power of flight completely. So while the species is completely unknown only a few miles from its moorland haunts, it does enjoy the most extraordinary acceleration. Though it cannot fly at 60 mph, its 0–60 figure, to use the motoring metaphor, is astounding. Like a drag-racer, it just cannot keep it up for very long. Such an explosion of speed may enable it to outstrip a predator, but the bird is no use at all in a prolonged chase. The wings of a Red Grouse are short and rounded, giving enormous lift and acceleration, in sharp contrast to the long, narrow and pointed wings of the Arctic Tern which spends a huge proportion of its life in the air.

Although one can learn a great deal about the type of life that a bird lives by its structure, a short, rounded wing does not necessarily indicate a residential, "Red Grouse" style of living. The Wren, or Winter Wren as it is known in North America, is a long-distance migrant that breeds in the boreal zone of the northern conifer forests and flies southward almost to the Mexican border to winter. Over most of the densely populated regions of that continent, therefore, it lives up to its name of "Winter" Wren. Yet this same little bird, with tiny rounded wings, is a resident virtually throughout its European range. British Wrens, for example, rarely cross the Channel to France. This lack of mobility has led to distinct subspecies evolving on the islands of the eastern Atlantic – for example, in the Faeroes, Shetlands, Outer Hebrides and even on isolated St. Kilda and Fair Isle.

The Black-headed Heron is the typical large heron of Africa south of the Sahara. Though closely related to the similar Madagascar Heron, which is an inhabitant of mangroves, the Black-headed Heron is the only member of the genus Ardea *to forage predominantly on dry land, where it feeds on mammals, frogs and insects found among grasslands.*

Yet, just as the Wren in Europe is largely resident, so it is along the west coast of North America. Here, too, distinctive subspecies have evolved in such locations as Kodiak Island, Semidi Island and the Aleutian Islands of Alaska. Also, at the westernmost end of the Aleutian chain there is a separate subspecies on Attu and Agattu Islands. It is interesting to speculate that, being the only wren to occur outside of the Americas, the Wren colonized Eurasia (perhaps as the House Wren) and then evolved to become a separate species that reinvaded America as the Winter Wren.

Birds do then vary in size and shape, in their numbers and in their flying abilities, in the movements that they undertake each year and in their breeding routines and nest sites. We know that some birds lay only a single clutch of a single egg, and that others may produce several clutches of five or more eggs. In either case, we also know that the routine is appropriate to maintain the population at its ultimate level under normal circumstances. Perhaps it is appropriate here to cast an eye over "normal circumstances" for our world is becoming increasingly "abnormal".

Hurricanes, earthquakes, floods and volcanic eruptions are natural phenomena, albeit violent ones. Birds subject to such violence must inevitably suffer as the change is so rapid that they cannot adapt quickly enough. More gradual changes, however, can often be accommodated. The coming and going of the last ice age some 10,000 years ago was one such gradual process that birds were able to deal with. They changed their distribution, moving gradually southward as the ice shield came and northward again as the ice receded. In doing so, species, or at least subspecies, evolved in separate populations; but their behavior changed as well. The phenomenon of a European migrational divide is a prime example of the effects of recolonization as the ice receded. Such a system is evident, for example, in the White Stork and the Black Kite, both of which recolonized from two distinct directions to meet in central Europe. Even today the two species divide in that area with western populations moving southwest through Gibraltar and eastern ones southeast over the Bosporus.

Somewhere between the lengthy time scale of an ice age and that of an instant volcanic eruption, lie the changes that man has inflicted and continues to wreak on the planet. It is axiomatic that the pace of change is increasing and that birds will find it progressively more difficult to adapt as the speed of change moves from the slowness of the coming and going of an ice age to a more explosive eruption. Today there are progressively fewer parts of the world that are not subject to human manipulation. The rainforests, the richest of all habitats in terms of species numbers, are being destroyed with amazing disregard. Grasslands are disappearing under the plow, water tables are being pumped dry, the seas are being overfished and even Antarctica is threatened with mineral exploitation.

Will there be, we could reasonably ask, any place for birds in this brave new overexploited world? The answer is, of course, yes. There will always be a place for those species that prosper by living alongside man. So, we can expect more sparrows, weavers and starlings, but can we also hope for stable populations of albatrosses, terns, macaws and eagles? The answer to this question is quite definite – no! Regardless of how enterprising, how resourceful, how determined and how expert we are, we *cannot* maintain a species that is dependent for its existence on a habitat that we as humans are hellbent on destroying.

To my mind, only control of our own population will save this planet from destruction, and that means that human birth control is the single and most important factor in the conservation equation. Put in its simplest form, "more people" equals "less birds".

I have delved in great detail into bird conservation in the final chapter of this book. Meanwhile, between this message of gloom and the despondency of the epilog – which is not compulsory reading – let us celebrate the living birds of the world in all their beauty and variety.

CHAPTER THREE
EURASIA
THE PALEARCTIC REGION

The Lanner is a middle-weight falcon, somewhere between a Merlin and a Peregrine in size and power. Essentially a Palearctic species it nevertheless roams as far as Africa and the Orient.

Opposite: *A pair of Black Storks perform a bill clapping greeting ceremony at their Czechoslovakian tree nest. Though, like their close relative the White Stork, they are summer visitors to the Palearctic, they do not share that species' confidence with humans. They thus hide away among dense forests and on rocky crags where they are seldom disturbed.*

The Palearctic region extends from the Canary Islands and Iceland eastwards across Europe and Siberia to China and the Bering Straits; or, if you wish, from the North Atlantic to the North Pacific. To north, east and west its borders coincide with the sea; to the south they are less clear-cut. The great Sahara desert is clearly a divide between the Palearctic and the Ethiopian region of Africa, but where exactly one ends and the other begins has been a source of debate among generations of zoologists. All, however, agree that the border between the Palearctic and Ethiopian regions lies somewhere across the middle of the Sahara.

The southern boundary continues across the Red Sea, crosses the southern part of the Arabian peninsula, and then turns northwards along the western shores of the Persian Gulf to join the great mountain ranges of southern Asia that include the Hindu Kush and the Himalayas, which, as we have noted, were formed when the plate of the Orient crashed into the Palearctic. The boundary, once again becoming less clear-cut, then extends

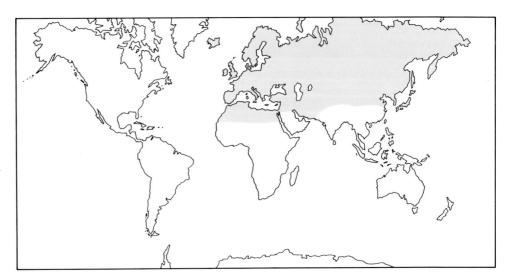

A Common Whitebait brings food to its young at a nest among ground vegetation. When drought struck the Sahel zone of Africa south of the Sahara in the 1970s, the Whitethroats which winter there suffered a catastrophic decline in numbers, the first sign of the misery that was to affect the human population of this region.

eastwards across China roughly along the course of the Hwang Ho River to the China Sea between Shanghai and Peking. Yet despite the difficulties involved in drawing this southern "land line" the Palearctic has more in common with the Nearctic, from which it is separated by the sea, than with the lands of the Ethiopian and Oriental regions with which it shares land boundaries.

A glance at a globe is sufficient to show that the world's great landmasses lie in the northern hemisphere. Yet these landmasses not only lie north of the equator, but also mainly north of the tropic of Cancer. This circumstance of geography means that the world's largest landmasses have a decidedly seasonal climate, showing a considerable difference between winter and summer temperatures and length of day. It is not surprising, therefore, that large numbers of birds have evolved a seasonal pattern that parallels these climate differences.

Palearctic birds breed during the northern summer and have evolved large clutches and a multibrood tactic to overcome this seasonality. Species that are resident in these latitudes may be able to find sufficient food to feed themselves during the colder and shorter days of winter, but are quite unable to raise young at that time. It should not be forgotten that even seed-eating birds tend to rear their young on a rich diet of insects and their larvae, food that is in short supply in winter. For similar reasons, insects breed in

A solitary Siberian Crane at its wintering grounds in Bharatpur, India. In 1972 over 50 of these Old World equivalents of the American Whooping Crane wintered there. By 1991 their numbers were down to only five. It can only be a short time before anyone wishing to see this splendid bird will be able to do so only in China.

To survive the northern winter the Willow Grouse feeds on the seeds and shoots of tundra vegetation and hides deep among drifted snow, igloo-fashion. Its feet are feathered for insulation and its brown plumage changes to white for the winter as a protection against predators.

summer because the vegetation on which their young depend is largely unavailable in winter. So, at the end of the chain, birds depend on a growth of vegetation which is available in summer, but not in winter.

Faced with a lack of winter food, birds have adopted one of two tactics: some change their diet to exist on the bountiful harvest that remains left over from the previous autumn; others move away to milder climes – in other words, they migrate.

The Palearctic is thus a summer home to huge numbers of migrants, birds that utilize the abundance of summer food and long hours of daylight in which to forage, but which then leave to seek food elsewhere during the hard, short days of winter. Some may make only short journeys to avoid land that is covered with snow and ice; birds such as ducks, geese and swans fall neatly into this category. Others may make huge migrations, taking them from the northern to the southern summer and back again each year. Many of these are true globe spanners like the Arctic Tern, Turnstone (Ruddy Turnstone in North America), and Bar-tailed Godwit, while others are more confined, with European Swallows going mostly to sub-Saharan Africa and Whitethroats to the Sahel zone. Even closely related species may exhibit quite different tactics, like two of the leaf warblers. One, the Chiffchaff, seldom migrates much further than the Mediterranean or the west coast of Africa, whereas the Willow Warbler is a trans-Saharan

migrant that flies this extraordinary barrier twice each year.

The geographical facts of Palearctic life, then, do make the region a primary home for migrants during the summer, and a considerable number of species have evolved to take advantage of the wealth of summer food. The waders and terns probably evolved somewhere in the north, though it is impossible to say whether this "somewhere" was in the Palearctic or Nearctic. Similarly, while it seems likely that the pheasants, owls, cranes, shrikes, tits and nuthatches did evolve in the Palearctic, they may well have had their origins elsewhere. Of all the world's 157 or so bird families, only the accentors can be said to have *probably* originated in the Palearctic.

The Hedge Accentor or Dunnock is confined to that part of the region west of the Urals and it ranges southwards as far as the Mediterranean and eastwards to Asia Minor and northern Iran. Its nearest known relative is the Japanese Accentor, which occupies the same latitudinal niche in the Pacific and has a similar song. The other ten accentors are found through the Palearctic, with the Alpine and Rufous-breasted Accentors penetrating as far south as the Himalayas and Radde's Accentor as far as the Yemen.

All accentors are ground-dwelling birds with fine, warbler-like bills ideally suited to picking insect food from among rocks and crevices. Most are found at altitude and have decidedly restricted ranges as a result. Only the European Hedge Accentor has managed to escape from a mountain

Largest and most powerful of all owls, the Eagle Owl is resident throughout its range. While food is plentiful in summer, it survives the winter by hunting mammals as big as hares. Like other large predators, it is capable of surviving lengthy periods without food.

Probably the most abundant of Palearctic summer visitors, the Willow Warbler is a tiny long-distance migrant that breeds from Europe right across Siberia. Yet even the most easterly birds head southwest to winter in Africa south of the Sahara, a huge transcontinental journey for such a small bird.

The distribution of the accentors shows that this small family is virtually confined to the Palearctic region, doubtless its original home. It also shows that, apart from the Hedge Accentor's occupation of Europe, these birds are found predominantly in mountain or tundra regions.

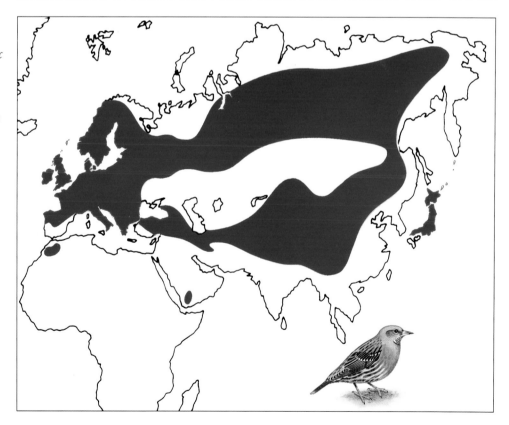

habitat and spread over large areas of lowland, though the Siberian Accentor occupies low-lying but alpine-type habitat among the tundra of northern Siberia. This environment, however, is quite different to the hedges, gardens and woodland margins preferred by the Hedge Accentor.

While the accentors may be the only truly endemic family to inhabit the Palearctic, there are certainly others that are predominantly Palearctic birds. The wildfowl, as we have seen, may have originated in the Nearctic, but some groups are predominantly of Eurasian distribution and there seems little reason to doubt their origins. The "gray" geese of the genus *Anser*, for example, are predominantly Palearctic birds, breeding among the summer-rich tundra and migrating to milder climes for the winter. Only the White-fronted Goose has managed to colonize North America in any positive way, but several others have spread as far west as Iceland. The Graylag Goose breeds in Iceland and the Pink-footed Goose has spread as far as eastern Greenland, yet both return to northwestern Europe in winter. The

Like most other members of its family, the Alpine Accentor is essentially a mountain bird that inhabits the sparse screes and scrub found only at high altitudes. It is a member of the only truly endemic Palearctic family, but has successfully colonized westwards as far as the Sierra Nevada of southern Spain, where it is commonly seen around ski hotels.

Above: *Although the Graylag Goose breeds as far west as Iceland, it nevertheless retains links with its Old World origins by migrating exclusively to western Europe in winter. It is generally accepted that colonizing birds follow their route of colonization on migration, thus providing solid evidence that the "gray" geese originated in the Palearctic region.*

It is notable that although the White-fronted Goose has colonised the New World on the west coast of Greenland, these birds still make a transatlantic journey to winter in Europe along the west coast of Britain and Ireland.

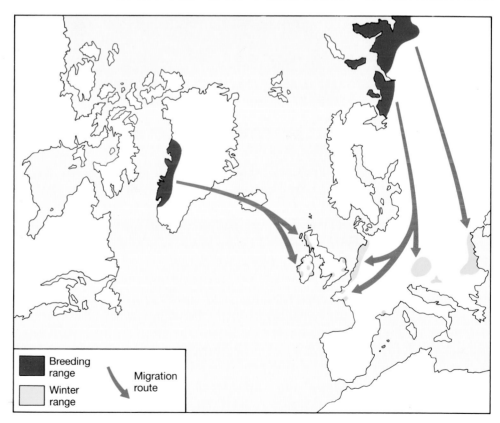

Breeding range

Winter range

Migration route

White-fronted Geese that breed in western Greenland, and perhaps even in the Canadian archipelago, still move eastwards to winter in Europe. Such movements are usually regarded as indicators of the route followed by the species during the process of colonization and, as with so many northern breeding species, the route would seem to follow that of the retreating ice cap after the last ice age. It is this factor that accounts for the migrational divide in central Europe, with some birds heading south-westwards and others to the southeast. In the case of soaring birds this can be explained by virtue of narrow sea-crossings in these directions, but even among small birds, with little fear of the sea, the divide is still apparent.

The large soaring birds that inhabit the Palearctic fall into two main groups, birds of prey and waterbirds. As long ago as the ancient Greeks, the regular comings and goings of these birds were noted and remarked upon. Some, like the White Stork, were regarded as signs of spring and as a talisman of good fortune, as bringers of babies. Certainly, the great swirling masses that follow the main migration routes are spectacular and among the most dramatic wildlife sights to be seen in Europe.

Like the birds of prey, storks rely on warm upcurrents of air to gain lift and then, by gliding, to cover distance. On migration they move from one updraft, or thermal, to the next until they arrive at the sea. Thermals, as any glider pilot will explain, do not develop over the sea and soaring birds must therefore use the narrowest of crossings. Through most of the Palearctic there are no seas to act as barriers, though even large lakes are avoided by these birds. In the west of the region, however, lies the almost landlocked Mediterranean that forces soaring birds to cross at its narrowest points – at Gibraltar and the Bosporus. Here, in both spring and autumn, great flocks of soaring birds gain height over the adjacent landmasses before gliding and flapping their way over the narrow straits. It is a splendid spectacle and one that attracts watchers from all over Europe. In the east the funneling effect is enhanced by Russian birds avoiding the Black Sea by concentrating along both its eastern and western shores. While as many as 400 eagles may fly over the Bosporus in a single day, these numbers are dwarfed by the number of White Storks that swirl overhead in their thousands.

These birds continue through Asia Minor and concentrate again along the eastern shores of the Mediterranean, where they are joined by birds that have taken the eastern route around the Black Sea. Here the route follows the Rift Valley and part of the Dead Sea, before crossing the Red Sea into Africa at Sinai. Though it has been little studied the importance of the Rift as a migration route for soaring birds is certainly apparent in Ethiopia and in Kenya where great flocks of raptors may be seen in autumn. Mostly they fly from thermal to thermal overhead but, on occasion, exhausted birds may take shelter from the sun beneath a desert bush, perhaps to die. It is a strange thought that a Common Buzzard, bred in the foothills of the Urals, may eventually die under an acacia in southern Ethiopia. Nonetheless, for the species, migration has an advantage that outweighs the death of an individual.

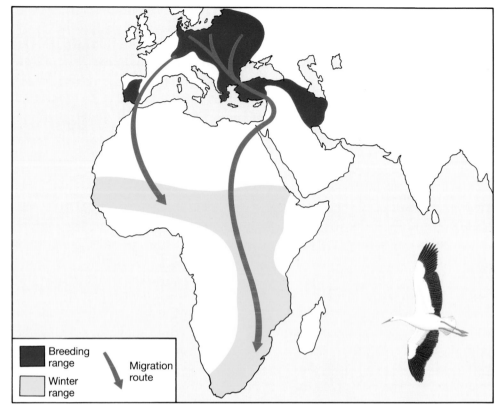

The summer and winter ranges of the White Stock are separated by the Mediterranean Sea, which forms a significant barrier to birds, such as these, which rely on soaring rather than flapping flight. The White Stork thus crosses at the narrowest points – the Strait of Gibraltar and the Bosphorus.

Breeding range

Winter range

Migration route

Traditionally regarded as harbingers of spring as well as a talisman of fertility and good fortune, White Storks, seen here over the Turkish Bosphorus, are long-distance migrants that utilize the warm rising air of thermals to soar effortlessly over many miles. As thermals are only generated over land, the birds concentrate at the narrowest of sea crossings such as here and at the Strait of Gibraltar.

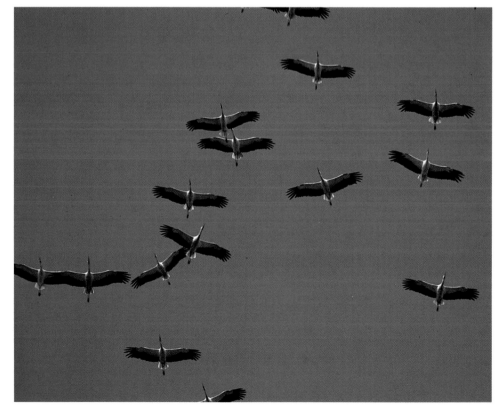

A dark phase Eleonora's Falcon at its cliff nest on a Mediterranean island. By timing its breeding season to coincide with the autumn migration of small birds between Europe and Africa, this fast and agile hunter is able to exploit an abundant source of food with which to feed its young. As the food supply passes through its island haunts, it has no need of territory and is colonial as a result. Further, it copes with variations in supply by storing surplus food in "larders", or food dumps.

Awareness of the traditional migration routes of the soaring birds was, quite understandably, applied to other birds, and for centuries it was widely thought that migrating birds of all shapes and sizes followed clear-cut routes between their breeding zones and their winter quarters. Such a misconception lasted at least until the middle of the twentieth century and was only finally dispelled with the advent of radar. Early versions of these remarkable electronic wonders showed all manner of echoes, including migrating birds, and it soon became obvious that the vast majority of small birds migrated on a broad front. It was, perhaps, sheer chance that the original studies of birds by radar were undertaken in Europe, in the western part of the Palearctic. For it was here that the most obvious migrational divide had been observed, a divide that was easily explained by the presence, to the south, of both the Mediterranean and the world's largest desert, the Sahara. Before radar it was quite reasonable to suppose that migrating birds were travelling in different directions to avoid these formidable barriers. After radar, ornithologists recognized that even small

Kittiwakes construct their bulky seaweed nests on tiny ledges of the most fearsome cliffs, forming some of the largest and most spectacular seabird colonies of the Palearctic. Unlike most other gulls, outside the breeding season they may roam the northern oceans, but in summer they must forage within range of their homes, though they may cover long distances to favored fishing grounds.

warblers, weighing little more than half an ounce, were crossing both barriers in a 40-hour nonstop flight.

In fact, it should have been obvious before images on the radar screen made trans-Mediterranean–Saharan flights incontrovertible. The many small islands in the Mediterranean are the home of Eleonora's Falcons, gregarious hunters that breed late in the year when the abundance of small migrants offers a bountiful source of food on which to feed their young. If there were eastern and western routes followed by these migrants then the Falcons would doubtless be concentrated at each end of the sea. As it is, they are spread right through the Mediterranean indicating that migrants cross virtually from one end to the other.

The breeding routine of Eleonora's Falcons is quite astonishing, for not only do the birds breed in colonies and switch from an insectivorous to an avian diet in the autumn, but they also hunt after dark and create "larders" for storing excess food. Though such behavior is unusual it is not unique, even within the Palearctic. The Sooty Falcon breeds among the jebels and stony wastes of the Sahara and the deserts of the Middle East, and it, too, is dependent on autumn migrants with which to feed its young.

Not all the birds that inhabit even the most northern regions of the Palearctic are, however, migrants. Some, by virtue of their habitat, may find food all the year round. Perhaps the most obvious of these are the seabirds, such as the gulls and the auks. It is generally recognized that the auks probably had their origins in the northern Pacific, in the region of the

Although the relatively weak flying powers of the Common Guillemot confines it to finding food near its summer breeding colonies, in winter these birds spend their lives at sea, often out of sight of land. Such dispersal is quite different to true migration, being more akin to the feeding flights of the related gulls than, for example, the migrations of terns.

Bering Straits, but some have managed to colonize the North Atlantic and develop as separate species.

The auks are able to find food wherever the sea remains unfrozen and it is no accident that they abound along the North Atlantic and northern Pacific coasts of the Palearctic. Both regions are blessed with food-rich seas and abound with fish. In their thousands, sometimes tens of thousands, they pack their traditional nesting cliffs. Common and Brunnich's Guillemots (Thick-billed Murve in North America) nest on tiny ledges, where their single egg is specially shaped to spin rather than roll. But the most numerous bird of all, the Little Auk or Dovekie, breeds the farthest north, among the rocky screes of the world's most inhospitable Arctic islands. Colonies of these tiny Starling-sized auks may number millions of birds, all dependent on the flush of life that blooms in the Arctic seas.

Auks, like the southern penguins, use their wings for propulsion underwater, literally "flying" beneath the waves. For them wing structure is thus a compromise between flying and swimming. One species, the now

An aquatic jewel, the European Kingfisher pauses for its portrait with a fresh fish for its nearby young. This is one of only a handful of highly colored birds to have colonized the temperate northern hemisphere – a region where dully-colored birds clearly have an advantage.

To compensate for their dull, camouflaged appearance, many northern birds, such as these Nightingales, have developed particularly fine songs to attract a mate and defend their territory. In the case of this species, the song is frequently produced after dark when most other birds are soundly sleeping – a fact that only adds to their reputation as arguably the world's greatest songsters.

extinct Great Auk, dispensed with the power of flight completely. It was exterminated in its few known Atlantic colonies by direct human persecution. Today, all of the auks face disaster at the hands of man. These birds are the prime victims of oil pollution and of overfishing, both of which may eventually reduce their populations to nonviable levels.

Dramatically colored birds are poorly represented in the Palearctic, as well as in the Nearctic come to that. Such species tend to develop only in areas where being brightly colored has a distinct advantage. Perhaps the most obvious of these areas are found among the vast tropical rainforests, where bright coloring has an advantage in defending a territory and attracting a mate, but where the disadvantage of being conspicuous, and consequently obvious to predators, does not apply. The Palearctic is, therefore, decidedly short of parrots and sunbirds and totally lacks many colorful families such as toucans, motmots, turacos and hummingbirds. Such colorful families that do occur in the Palearctic are represented by only one or two species that have managed to colonize what would appear to be unsuitable areas. Thus, of the world's 90 species of kingfishers only six can be found in the Palearctic, and two of these are black and white rather than the bejewelled gems that we associate with this family. Furthermore, of the world's 23 bee-eaters, just four have penetrated the region and two of these have done so only marginally.

Although the titmice are a widely distributed family, the fact that a majority of species inhabit the Palearctic leads us to postulate that they have their origins in this region. The highly successful Blue Tit has managed to colonize suburban gardens and is now as abundant in this new habitat as in its more natural woodland home. An ability to live alongside humans is clearly an advantage as mankind expands its occupancy of the planet.

In contrast, the Palearctic is well endowed with brown and buff birds that spend much of their time in self-effacing pursuits. Some, like the famous Nightingale, may be excellent songsters, but there is no monopoly of song among the dully colored. Nevertheless, these apparently similar, small, dull birds do pose interesting problems and it is significant that the largest population of bird-watchers in the world has developed in Europe where such birds abound. Of course, there are dull, brown birds everywhere, but the Palearctic warblers are in a class of their own, rivalled only by the Afro-Asian cisticolas when it comes to identification.

The warblers of the genus *Phylloscopus* probably originated in the Sino-Himalayan region. There are some 38 of these so-called leaf warblers and, virtually without exception, they are difficult to identify. In fact, they are so difficult that most European birders consider them a great challenge and the most severe test of identification skill. They have thus become a sort of litmus paper of one's standing – but they are still regularly misidentified even by those who should know better. These tiny little birds also have the endearing habit of vagrancy – of turning up hundreds, even thousands of miles from where they should be. Thus a Siberian waif like Pallas' Warbler regularly occurs each year in Britain, to the great delight of anyone fortunate enough to see it. Such vagrancy is difficult to explain and has led to heated debate among generations of ornithologists. Complex weather maps are often produced in evidence, but the most reasonable explanation is

that young birds sometimes get their migrational directions completely
wrong and head off in exactly the reverse direction to the one they should
follow. This phenomenon is appropriately referred to as "reverse migration".

Of all the leaf warblers, perhaps the most notable is the Arctic Warbler.
Arctic Warblers spend the winter in southeast Asia, but while some then fly
northwestwards as far as Swedish Lapland, others head northeast to Alaska.
Almost as dramatic is the similar Greenish Warbler, a bird that is abundant
in winter among the Indian plains, but which breeds as far west as Poland.
Unlike the other *Phylloscopus* warblers that have colonized Europe, these
two species remain loyal to their region of origin, for both Willow and
Wood Warblers have become trans-Saharan migrants that winter in Africa.

There are several other bird families that can with some justification be
regarded as typically Palearctic birds. Two, the tits and the true finches, are
among the most familiar and best-loved groups of birds in the world. Both
take readily to gardens and both come regularly to feeders, making them a
joy to bird gardeners wherever they occur. Perhaps the most familiar in this
respect is the Blue Tit, which ranges throughout Europe as far as the Urals
but no further. Distribution of the Coal and Great Tits, in contrast, extends
eastwards as far as Kamchatka and the Bering Sea. The Willow Tit, too,
extends right across the Palearctic, whereas the Marsh Tit is found only in
Europe and again in far eastern Siberia, China and Japan. The tits are
essentially woodland birds that are mostly resident. In central Siberia they
are able to survive the coldest of temperatures, with many weeks below
freezing during the long winters. That they can do so is dependent on the
food supply available during this lean season in the forests.

Among the most strictly resident of all small temperate zone birds is the
Crested Tit. Largely confined to conifer forests in Europe, this attractive
little tit spends its entire life within its natal forest. Birds that breed in
Scotland, for example, were virtually confined to the native Scots pine
woods of the Spey valley before twentieth-century plantations offered it an
alternative home. Even then the Crested Tit was not prepared to cross open
ground to reach the new habitat. Today it has failed even to cross the
Cairngorm Mountains from the Spey to the Dee valley, a distance of no
more than ten miles of nonconifered hills. Such resistance to movement is
unusual among birds, and particularly among small, temperate passerines.

The nuthatches are also a remarkably sedentary group of small birds – so
much so that while the Palearctic is largely occupied by a single widespread
species, the Common Nuthatch, other nuthatches have developed in well-
defined and often quite small areas. For example, the Kabylie Nuthatch is
found only in a small area of mountain forest in northern Algeria; the
Corsican Nuthatch is found only on the island of Corsica; Kruper's
Nuthatch is a native of the spruce forests of the mountains of Turkey and
the Caucasus; the Yunnan Nuthatch is restricted to southern China; and the
White-cheeked Nuthatch occurs in two areas at either end of the Himalayas.
The propensity to speciate in isolated areas of habitat is a characteristic of all
birds, but it is particularly obvious in birds that, like the nuthatches, are
largely resident.

Although large fish-eating eagles can be found in many parts of the world, only the Palearctic can boast four distinct species. In the Ethiopian region there is the African Fish Eagle, replaced in Madagascar by the Madagascar Fish Eagle; in North America there is the Bald Eagle, while Australia has the White-breasted Sea Eagle, which also ranges northwards as far as India, Burma and Vietnam, thus clipping the Palearctic. Sanford's Eagle is found among the islands of the Pacific, leaving three species found exclusively in the Palearctic to complete the genus.

Most widespread is the White-tailed Sea Eagle, which ranges from Scotland, where it has been successfully reintroduced, to the furthest part of Siberia. Here it overlaps with the huge Stellar's Sea Eagle around the Sea of Okhotsk. In the south and central areas the smaller Pallas' Sea Eagle breeds from the eastern Caspian south to the plains of northern India. Although all of these birds are called "sea eagle", none are confined to the sea. Most breed inland beside lakes and marshes and, although capable of fishing by plunge diving, they are also notable scavengers. No doubt the first-named species, the White-tailed Sea Eagle, passed its name to the others after being named in western Europe, where both Scottish and Scandinavian birds are residents of sea coasts. Elsewhere in Europe this same bird inhabits lakes that, as throughout Siberia, freeze solid in winter and prevent the birds from feeding. Thus most of these eagles are at least partial migrants.

Sadly, these are often persecuted birds and in several parts of their range bounties were formerly paid for their destruction. A more enlightened attitude in recent years has seen such ÿstems abandoned and replaced by active conservation measures. In Norway, for example, the number of White-tailed Sea Eagles has grown to over 400 pairs since a bounty system was abandoned, and it is this population that has been used in the Scottish reintroduction scheme. Elsewhere, however, the decline in numbers continues.

Other raptors have fared little better, partly as a result of habitat destruction but also by direct persecution. Nowhere, perhaps, is this more apparent than in the Mediterranean where virtually anything that flies is considered fair game. In Sicily, for example, the regular autumn passage of Honey Buzzards is pursued with undiminished enthusiasm, despite all the endeavours of Italian conservationists. In Britain birds of prey are still illegally shot by gamekeepers, though the number of birds killed has dropped considerably in recent years. In other areas the destruction continues unabated and gibbets hung with corpses are still to be found. The killing may be based in part on some misconceived idea of protecting populations of game birds – for example, Red Grouse in Britain and Red-legged Partridges in Spain. But much of the slaughter is quite indiscriminate and takes a toll of birds that in no way interfere with game birds.

In parts of the Palearctic many birds of prey have been totally exterminated. In western Europe, for instance, the range of the Golden Eagle is fragmented, with birds largely confined to remote mountain areas. Even in the most conservation-conscious areas, these magnificent birds are killed by shooting, trapping and poisoning, and in some places egg collecting

still takes a serious toll. For conservationists it all seems very daunting; even when they appear to be winning the battle new dangers can appear.

During the 1970s and 1980s, for example, the newly oil-rich populations of the Middle East became an expanding market for falcons. Once regarded as the preserve of rulers, the newfound wealth of even minor officials and businessmen was dedicated to the traditional sport of falconry. The result was that the price of a falcon rose dramatically and many saw the opportunity for easy profit. With the supply of passage Lanners and Sakers drying up, attention was turned to the strongest populations of Peregrines in Europe around the coasts of Britain. Strangely, the falcon stealers quickly discovered that they had to guard the nests they intended to rob from egg collectors. So, in the early stages of the breeding cycle the robbers became the protectors. Unfortunately, this trade in young Peregrines still continues and the sums of money involved remain significant. While it is impossible for conservationists to guard each and every nest, nevertheless their

A Common, or Eurasian, Nuthatch feeds its blind and naked chicks inside its protective tree-hole nest site. Mostly nuthatches are resident wherever they are found, a factor that has led to the development of isolated species in several parts of the western Palearctic. The need to utilize mud while nest building may lead the Common Nuthatch to plaster up the entrance hole to its nest, even when it is already the correct size.

Opposite: *Decimated by human persecution, then by poisonous pesticides the Peregrine Falcon now faces the threat of nest robbing by falconers, as this old established sport becomes a status symbol among the* nouveau riche *of the Middle east. Only the efforts of dedicated watchers backed by well organized conservation bodies can stem the tide of thieving that threatens these birds throughout their range.*

The beautifully marked Impeyan, or Himalayan Monal Pheasant, is typical of this group of colorful, but secretive, Palearctic birds. Though several species are found outside its boundaries, it seems highly probable that these birds originated in the forests of northern China. Today, some species are more widespread following their introduction as the world's primary sporting quarry.

activities are a considerable deterrent and falcon robbers are arrested and punished every year. That an active protection policy can work is shown by the continued expansion of the range of these falcons around British coasts.

Although the avifauna of the Palearctic region is poor in terms of the number of families that are found solely within its borders, there are, as we have seen, a good number that undoubtedly originated there and which have since spread to other regions. To the tits, nuthatches and warblers already mentioned, we can add the pheasants, shrikes, Old World flycatchers, creepers and crows, all of which probably developed in the region. There are good reasons to believe that the cranes, too, evolved within the Palearctic. This family consists of only 14 species, but they are among the most beautiful and spectacular of all birds. Two species are found in North America, three are confined to Africa and one each to Australia and India; all the rest are found within the Palearctic.

Cranes are large and gregarious birds, some of which perform lengthy migrations in spectacularly noisy flocks. In general they are a declining family and several species are in serious danger of extinction. In the Palearctic the Common Crane still occurs in many areas in large flocks. It

This portrait of a dark phase Arctic Skua on its tundra breeding grounds gives no hint of the piratical lifestyle of this powerful predatory seabird. Widespread in both Palearctic and Nearctic regions, these birds range southwards through the world's oceans in winter, in their search for gulls and terns to rob and plunder.

breeds among temperate marshes and tundra from Scandinavia right across Siberia and spends the winter as far south as North Africa, the Middle East and India. Its movements across western Europe follow a well-defined and traditional route that extends from the Baltic across Germany and southern Holland to France, where the creation of new reservoirs in the Marne region has encouraged a significant number of these birds to winter far to the north of their normal range at this season. Most, however, continue beyond the Pyrennees with the bulk of the population centered in northern and central Spain. In spring up to 5,000 cranes gather in southern Sweden at Lake Hornborgasjon, courting and dancing while they await the thaw of their nesting grounds to the north. In contrast the population of the Japanese or Manchurian Crane is so small that the species has been on the brink of extinction for several generations.

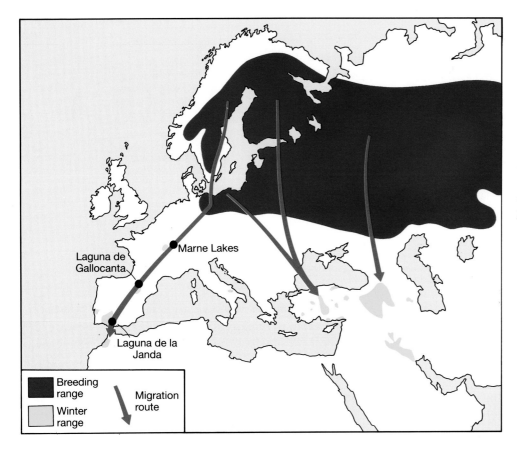

Laguna de
Gallocanta

Marne Lakes

Laguna de la
Janda

Breeding range
Winter range
Migration route

Common Cranes divide into western and eastern populations in central Europe. The well studied western group follows a narrow corridor from the southern Baltic to its Spanish winter quarters, with traditional stop-over points along the way. The creation of the large reservoirs known as the Marne Lakes, has offered these birds the opportunity to winter much further to the north than previously.

The beautiful Siberian Crane has also often appeared to be tottering on the edge of disaster. Breeding in three widely separated parts of Siberia and nothern China and with three quite distinct migration routes to separate wintering grounds, this large white crane has steadily declined in its two most westerly wintering areas. In Iran it has now all but disappeared from the marshes of the southern Caspian. In India the sole population at Bharatpur near Agra has declined to less than 30 birds. Only in China is there still a healthy winter population drawn from its most easterly breeding grounds of Yakutsk in northern Siberia and the Hulun Nur region of Heilungkiang in China. The reasons for this decline have been variously attributed, but an increasing human population making ever greater demands on the land of southern Asia is, perhaps, the most likely cause. At the well-studied wintering site in India, Siberian Cranes are forced to feed on tubers in deep water rather than in the more favorable shallows because of competition with the larger and more powerful resident Sarus Crane.

It is somewhat ironic that the region with the most impoverished avifauna should be both the birthplace of ornithology and have the largest number of active bird-watchers. The Palearctic, however, is poor only in the number of species it supports. In terms of numbers of individuals it is possibly as rich as any other region of the world and is blessed with some of the world's most dramatic bird spectacles.

CHAPTER FOUR

NORTH AMERICA

THE NEARCTIC REGION

Bald Eagle, symbol of a nation and the first effigy of a bird in space. Despite its human significance, this magnificent bird of prey has been eliminated from much of its natural range by direct persecution. There are now huge eagle-free-zones over much of its native Nearctic range.

Opposite: *An Eastern Bluebird arrives at its nest hole with food for its young. Once one of the more typical birds of North America, the introduction of alien species that compete for nest holes has seen a decline in the population of a species that would grace any of the world's avifaunas.*

The Nearctic consists of the nations of Canada, the United States and the larger part of Mexico. It is easily delineated to north, east and west by the presence of large areas of ocean, though in the south the land boundary with the Neotropical region is only a little more complex. In origin North America formed part of Laurasia and remained joined to northern Europe until at least 65 million years ago. By this time birds had been in existence for 75 million years and it is, therefore, not surprising that the two continents share so many avian characteristics. Indeed, the two are often treated as a super-region, the Holarctic, by some authorities. Nevertheless, as we shall see, there are significant differences between the birds of these two great northern regions and separate treatment follows the majority view favored by zoogeographers.

In the north the Nearctic extends as far as the land itself and is contentious only in including Greenland, which, surprisingly enough because of its proximity to America, shows considerable affinities with Europe. The

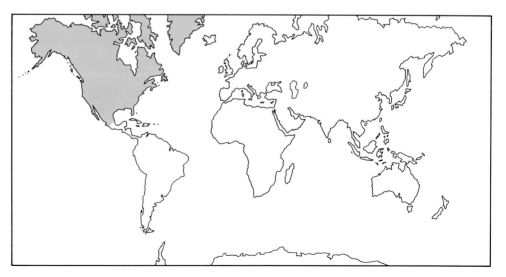

Confusing English bird names litter ornithological literature. This Nearctic bird is, quite reasonably, called White Ibis. But so too is the black and white ibis of the Oriental region which is, in fact, sometimes regarded as no more than a subspecies of the Ethiopian region's Sacred Ibis, but which current theory regards as a separate species. To add to the confusion there is also an Australian White Ibis which, quite naturally, Australians refer to simply as White Ibis. Surely it is not beyond the wit of mankind to resolve such confusion?

boundary thus passes between Greenland in the Nearctic and Iceland in the Palearctic. In the east the Atlantic is a clear boundary, while the Pacific performs the same function in the west. Here, however, the division between Asia and Alaska in the Bering Straits runs through the Aleutian chain of islands dividing what are parts of the United States between the Old and New Worlds. The southern boundary is more complex, but is drawn at the northern limit of the tropical rainforests, between Veracruz on the Gulf of Mexico and Guadalajara near the Pacific, or approximately the latitude of Mexico City. The southern sea boundary excludes the islands of the West Indies, which form part of the Neotropical region, but includes the Bahamas. It thus runs through the Straits of Florida which separate the United States from Cuba. Although the land link with South America offers obvious opportunities for an interchange of land birds between the two regions, it must not be forgotten that the two continents were quite separate until some 50 million years ago. At that time there were considerable gaps in Nicaragua and Panama that formed barriers to a full interchange of species. Nevertheless, the land links with the Neotropical in the south and the Palearctic across the narrow Bering Straits have both played a significant role in shaping the Nearctic avifauna, offering as they do an easy corridor for colonization. Parrots, for example, are generally regarded as being of Old World origin and as having colonized the New World via the Bering Straits. The fact that North America boasts only one species, the now extinct Carolina Parakeet, while South America is rich in species, shows that present distribution may easily disguise a group's origins.

The links with the Palearctic region are particularly apparent among larger birds. Thus, while only 16 Old World passerines can be found in North America, no fewer than 102 non-passerines are found there. The links between these two great northern continents are, however, more obvious at generic level, since although particular Old World species may be absent, they are often replaced, by similar and closely related species. In consequence, of the 329 Old World genera, no less than 114 also occur in America.

The northern continents also have a great deal in common with regard to landform, climate and natural vegetation. Thus, for example, birds that spend the summer among the great marshes of the Siberian tundra have their equivalents among the tundra of northern Canada. Some, like the Arctic Tern, are the same species, as indeed are the skuas of Eurasia and the jaegers of North America. Others are clearly closely related equivalents, like the Eurasian Greenshank and American Greater Yellowlegs, the Common and Spotted Sandpipers, the Hudsonian and Black-tailed Godwits and so on.

There are, in fact, five families that breed only in the joint Nearctic–Palearctic region, the Holarctic. One, the accentors, is, as we have seen, confined to the Palearctic and another, the auks, is shared by having its origins in the region of the Bering Straits. Three other families can be regarded as being of Nearctic origin, though none is endemic. The families of loons, grouse and phalaropes are all represented by more species in the Nearctic than in the Palearctic, though two, the loons and phalaropes, by only one extra species. All five of the world's loons (divers), breed in North America. (In order to best suit the subject matter, the reader will note that for this chapter and this chapter alone, I have used American bird names with any alternative names in parenthesis.)

The Common Loon (Great Northern Diver), is widespread throughout Canada as far south as the United States border and its loud, wild cries are well-known and well-loved sounds of the north. It breeds on ponds and lakes throughout the great boreal forests and northwards to well beyond the tree line. Although it is a great fish eater it is not regarded as a competitor by anglers, perhaps because there is so much fishing in these areas. Capable of diving for over a minute at a time, these birds lose their fine breeding plumage in winter and become rather grey and dull soon after they arrive at the coasts where they spend the hard months of the year.

The similarity between the Greater Yellowlegs (right) *and the Greenshank* (opposite) *is clear for all to see. The birds are structurally similar and share the same kind of habits and habitats and yet are found in separate zoogeographical regions. Such similarities between species may, as in this case, be due to a close relationship, but may equally be due to the process of convergent evolution.*

The other four species of loon all breed considerably further north than the Common Loon, and the Yellow-billed Loon (White-billed Diver) is confined to the most barren parts of the northern tundra. Though superficially similar to the Common Loon in both summer and winter plumages, the Yellow-billed Loon is a decidedly scarce bird that winters only along the Pacific coast as far south as central California, but which is rare south of the Canadian border.

All three of the world's phalaropes breed in the Nearctic, but while the two northern species, the Gray and the Red-necked Phalaropes, enjoy completely circumpolar distributions that take them to Siberia as well as Canada and Alaska, the larger, less aquatic, Wilson's Phalarope is exclusively an American bird. Phalaropes are waders that have taken to swimming in a very positive way. Typically, they spend their summers among marshy pools and shallow lakeside margins and their winters out of sight of land among the world's oceans. In both areas they feed on tiny food items – insect larvae in the summer and small planktonic creatures in winter. They also have the endearing habit of spinning on the water's

surface to create an upwelling of material including food from the marsh bed. Their buoyancy is legendary, but it is still incredible to see them bobbing up and down like corks among the great waves of the oceans.

Of the three families that are regarded as being of Nearctic origin, none is as diverse or as spectacular as the grouse. These medium to large-sized birds have adapted to a wide range of North American habitats and, until the arrival of man the hunter, were both widespread and abundant. Some species are still numerous, but others have suffered severely from overhunting and become locally extinct. The Heath Hen, a subspecies of the Greater Prairie-chicken, was wiped out completely from its range in the eastern United States by 1932. Other subspecies of this once widespread grassland bird merit a place in the *Red Data Book* which details birds in danger. Attwater's Prairie-chicken, the local Texan subspecies, was down to only a few hundred 20 years ago and other forms are little better off. The Lesser Prairie-chicken, regarded by some as yet another subspecies but accorded full specific status by the American Ornithologists' Union, occurs only in the southwestern United States and is declining fast.

Like many other grouse, the prairie-chickens are gregarious species that form special leks, or mating grounds, at which the males stake out tiny territories and perform elaborate displays to attract visiting females. During these jousting tournaments males raise and spread their wings and tails and expose colorful inflatable sacs at the sides of their neck. These sacs are also used to amplify a far-carrying booming sound that accompanies display. Charges, jumps and actual fights between males are commonplace and females are apparently impressed. Having mated the female takes full charge of the incubation and care of the young.

North American grouse occupy a wide range of habitats from grassland through dry, arid scrub to dense forests. One of the most prized specimens, at least by hunters, is the Ruffed Grouse, a forest-dwelling bird that is difficult to shoot and splendid to eat. It is unusual in producing its "song" by beating its wings over a hollow log rather than vocally. The drumming

produced is so vibrant that those fortunate enough to have enjoyed a close approach to a displaying bird may complain of painful ears.

Male grouse are able to make themselves highly conspicuous during the breeding season, but females are among the best camouflaged of all birds. Outstanding in this respect is the Sage Grouse of "cowboy film" country, which may sit immobile within a few feet of an intruder. These are large birds that are bitter to the taste because of their diet of sage shoots; even so, they are prized for their sporting characteristics.

If the divers, phalaropes and grouse are unique in being the Nearctic representatives of birds confined to the Holarctic super-region, they are far from being the only families of Nearctic origin. There are certainly five songbird families that have evolved in the region, but which have since spread southwards into the Neotropical region. The wrens are a typical example; many species are confined to South America, yet they are generally regarded as being of North American origin. In fact, there are considerably more genera and species north of the Panama "gap" than there are to the south. While wrens have evolved to occupy the most extraordinary range of habitats, elsewhere in the world they are all but absent. Of some 59 species, only one, the Winter Wren (simply, the Wren), has managed to colonize another continent. In Eurasia it is found from the

The Great Northern Diver, or Common Loon (opposite), is a Nearctic bird that has managed to colonize Iceland, its only Palearctic breeding ground. The White-billed Diver, or Yellow-billed Loon (above) lives farther north than its close relative, but has colonized westwards into northern Siberia. The two species are separated by a habitat preference in North America, but still maintain this preference even where competition is absent. In both continents the White-billed Diver is a scarce bird and a highly valued species for the watcher's list.

77

Opposite: *Breeding among the great boreal forests of Canada, the Winter Wren is an aptly named winter visitor to the most populated parts of North America. In Europe in contrast, where it is the only species of wren, it has dispensed with migratory habits and occupies a huge range of different habitats, from woods and gardens, to heaths and seacliffs. Being the only member of its family it is called quite simply "the Wren".*

A male Sea Grouse in full nuptial display in Colorado. The wealth of grouse species in the Nearctic region make it all but certain that this family of large, ground-dwelling birds originated in North America and colonized Eurasia via the Alaska–Siberia land bridge. As a family they are tenaciously territorial and sedentary making colonization via a lengthy flight extremely unlikely.

Pacific to the Atlantic and throughout this vast area it faces no competition from any other species of wren. As a result it has now moved into a wide range of habitats that, in its native land, are occupied by others. In Europe, for example, it is a bird of woodlands, gardens, scrub, marshes and even mountainsides and cliffs. Though at only four inches it is the smallest of the wrens, it is also the longest distance migrant. Winter Wrens from the Canadian boreal forests regularly winter as far south as the Gulf Coast of Texas and Louisiana, a journey of a thousand miles or more. In sharp contrast, North America's largest family member at over eight inches in length, the Cactus Wren, is resident in the southwestern United States and through Mexico. As its name implies, this is a desert species that constructs a nest of cholla cactus spines in which to breed and roost.

Among the best known of Nearctic birds are the mockingbirds of the family Mimidae, another group that have their origins north of Panama. Species include the Northern Mockingbird which is widespread in the United States and known for its renditions of the calls of other birds, along with less "natural" noises such as barking dogs, musical instruments and

A Cactus Wren arrives at its nest among the defensive spines of a cactus in the Sonora Desert, Mexico. A lining of fine grasses protects the eggs and young from the spines of which the nest is largely constructed. This bird, the largest of the world's wrens, is sedentary and often uses its nest as a safe roost throughout the year.

rusty door hinges. The Gray Catbird is similarly widespread and familiar, but while some birds are capable mimics, others are content with the cat-like "me-ew" after which the species is named. Among the most abundant members of this North American family, the various thrashers are a highly specialized group that exhibit an interesting propensity to speciate in the warmer parts of their range. The Brown and Sage Thrashers between them cover most of the United States, and a whole range of different species can be found in the southwest. Though superficially similar in shades of gray and with relatively long, decurved bills, each has a distinct habitat and feeding routine.

The birder traveling through the Nearctic region is inevitably impressed by two distinct bird groups – the wood warblers and the buntings – both of Nearctic origin. The fact that Americans call all of their buntings "sparrows" is not a deliberate attempt to confuse! These are a highly diverse group of small, mainly brown birds that occupy a wide range of habitats. In the Old World, buntings are usually birds of open ground that frequently choose quite obvious perches and song posts. In contrast, American buntings are found from dense woodland and thickets to equally dense aquatic vegetation. Many are great skulkers and extremely difficult to observe.

Among the most widespread and best known is the Song Sparrow,

Dramatically photographed underwater feeding on fish eggs, the American Dipper is one of only four dipper species in the world. Between them they cover several of the great landmasses, but are curiously absent from Africa and Australia. It is perhaps their need for a continuous flow of water throughout the year that accounts for their absence from the two southern continents that lack extensive, snow-holding mountain chains.

The Eastern Meadowlark breeds as far west as New Mexico, where its range overlaps with the Western Meadowlark. In this region, as elsewhere, the eastern species prefers damp fields and meadows, while its western cousin is found in dry fields. Such a division of habitats is essential to two species that are so closely related.

which breeds from the southern edge of the tundra southwards to northern Mexico. Through this vast range it occupies virtually any type of thicket and no less than 30 distinct subspecies have been described. Some of these are so variable in plumage and size as to appear to be different species, and it is difficult to understand that the large, grayish bird of the Alaskan Aleutian Islands is the same species as the neatly streaked brown bird of California.

Just as diverse are the various forms of the Seaside Sparrow that inhabit the tidal marshes of the east coast. So great is the difference between the subspecies of this bird that one, the Dusky Seaside Sparrow, was regarded until quite recently as a quite separate species. Found only in a restricted area of the Florida coast, this form has recently passed into extinction.

Despite being relatively common, some of these birds are decidedly – sometimes almost impossibly – difficult to see. Le Conte's Sparrow, for example, breeds in Canada and winters in the south-eastern United States, but is one of the most difficult of all American birds to locate. It seemingly spends almost its entire life crawling among the roots of dense grass where it is more likely to be identified as a mouse than a bird.

In sharp contrast, many of the wood warblers, the fifth and final passerine family of Nearctic origin (though now often included in the Emberizidae), are boldly marked, colorful birds. In spring, males in particular are patterned in shades of yellow, black and gray and are then awkward, but not impossible, to identify. In the autumn they lose their

bold patterns and become very similar. For this reason "fall warblers", as they are called in America, form one of the world's greatest identification challenges. Most wood warblers are summer visitors that spend the winter in Central and South America, though some do winter in the southern United States. Their migrations thus cover thousands of miles and "fall outs" of arriving migrants may be dramatic. Every spring North American birders concentrate at noted migration halts, such as the Gulf Coast and Point Pelee on the northern shores of Lake Ontario, to enjoy the sight of thousands of individuals of a huge variety of species stopping over on migration.

Between them the wood warblers occupy virtually every landform and habitat of North America, from the bogs and thickets of the subtundra to

Truly North American birds, a huge flock of Snow Geese lifts off from a field of corn that has been specially planted for these birds. In their length-of-the-continent migrations vast flocks are often dependent on man-made (and stocked) refuges to act as vital refuelling stations. Hunters may lie in wait nearby, but within the sanctuary the birds are safe.

The migration of the Snow Geese is divided relatively neatly between Atlantic, Central and Pacific flyways. Unlike most small birds, young geese have to be shown the route both southwards and northwards by their parents in fall and spring. There is thus a built-in propensity to follow a particular route, with particular stop-overs, within each breeding colony.

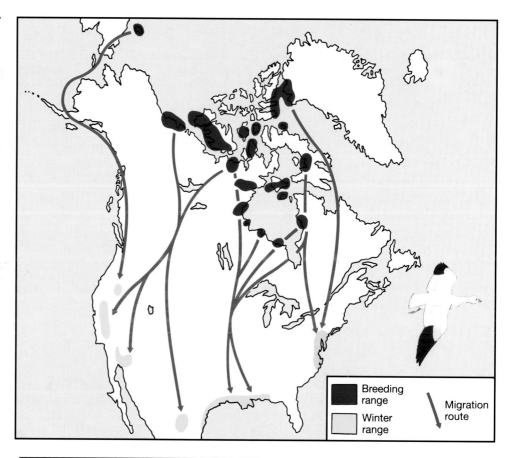

Breeding range

Winter range

Migration route

the cottonwoods that grow along dried-out desert watercourses. An exceptional adaption is shown by Kirtland's Warbler which exists only among young stands of jack pine. It is confined to a tiny area of Michigan where, to ensure its survival, the trees are planted and burned, according to a carefully designed management program, to provide the conditions it needs. Fortunately, not all wood warblers are so fussy.

If, then, the Nearctic region has some families of its own, even if they have spread to other regions, it also has families that are more widespread. Among the geese, for example, Brant and White-fronted Geese are also found in the Palearctic, but both Canada and Snow Geese are truly American. With their migration the length of the continent, these two species form one of the most spectacular wildlife sights in the world. Snow Geese in particular leave their tundra breeding grounds in the autumn and fly southwards in incredible numbers. They follow the three major flyways – the Pacific, central and Atlantic – and pause at traditional stopovers on the way to their wintering grounds. Here in their tens of thousands they feed voraciously, often on crops specially grown for them at refuges like Sand Lake on the central flyway.

Unlike the smaller passerine birds, geese do not have a built-in homing mechanism and on both the outward and inward legs of their first migrational flights young birds are led by the adults. Thus families keep

together through that first winter, and orphans either attach themselves to another family or are doomed. When Des and Jen Bartlett were filming *The Flight of the Snow Goose* for the British television series *Survival* they rescued several orphaned Snow Geese and became imprinted as foster parents. So while the other geese departed southward at the end of the breeding season, the Bartlett geese remained with their human "parents". Des and Jen took them southward by road in a trailer, but were prevented from importing them into the United States by customs officials. The solution was to release the young Snow Geese north of the border, drive across into the United States with an empty trailer and wait while the geese flew over by themselves, which they promptly did. Breaking no laws the family reloaded and motored on, courtesy of the US customs.

This leading of young by their parents is clearly essential, for elsewhere in the world where geese have been introduced they lose their migratory instincts and become residents. In Britain, Canada Geese sit around on ponds throughout the year. In North America, in contrast, they are in most cases as strongly migratory as the Snow Geese. These great comings and goings of vast flocks of wildfowl through the length and breadth of the Nearctic may be mirrored in the Palearctic, but they are nevertheless an integral and spectacular part of the region's avifauna.

There is nothing exceptional or unique about the Bald Eagle, national

Des and Jen Bartlett's "family" of Snow Geese that accompanied the Survival *camera crew on the journey from Hudson Bay to their home in the southern States. They paused here at Sand Lake, South Dakota, one of the most important stopovers for wild geese on the Central Flyway, to have their portrait taken in dramatic flight.*

symbol of the United States. It is a fish-eating eagle, closely related to similar species found in many other parts of the world. Once widespread throughout the continent, the Bald Eagle has been seriously reduced by direct persecution, by habitat destruction and by pollution. In fact, today the only really significant populations of this large brown bird, with the white head and tail typical of the genus, is in Pacific Canada and Alaska. Here the mountain wilderness, with its fish-rich rivers and lakes, offers perfect isolation for a really substantial population of these great birds.

Everyone has seen photographs and television films of Bald Eagles gorging themselves on the great runs of spawning salmon. Accompanied by grizzly bears and Glaucous-winged Gulls, the eagles stalk the rapids where great masses of fish negotiate the shallows. During the season, the living is easy and the birds have little to do but eat and sleep. Later, when the landscape is gripped by winter, the birds concentrate at fewer and fewer areas of open water. One such is the Chilkat Valley, where thermal springs keep part of the river free of ice. Spent salmon, deep-frozen higher up, drift downstream to the warm open water, offering an extended supply of food to several hundred Bald Eagles – the greatest, non-migrational concentration of eagles in the world. These birds are drawn to the Chilkat from an area of several thousand square miles, but by mid-winter they must disperse elsewhere as the food supply runs out. Some may resort to the coast, but others glean a living by scavenging the corpses of other animals that have failed to survive the hard season.

Once, the Bald Eagle could be found virtually everywhere in North America where rivers or lakes provided food. Gradually the accelerating pace of pesticide use invading their food chain, suburban development and industrialization reduced their numbers, and direct persecution added to the toll. Being at the top of the food chain, eagles are long-lived birds that take several years to reach maturity and, even then, lay few eggs and rear few young. They are therefore less able to recover from a disaster, and it will be a long, long time before they regain anything like their former numbers. Indeed, it seems highly unlikely that they will ever return to many areas where they were once numerous.

Because human colonization and industrialization is so recent we have an impressively well-documented history of the birds of the Nearctic. When the Pilgrim Fathers arrived the continent was rich in forests, lakes and marshes and all were full of wildlife. The Native Americans had taken some toll, but their traditional weapons enabled them to live happily side by side with wildlife. The coming of firearms changed all that and, with the frontiers-man's total disregard for, or ignorance of, the concept of game management, the destruction was enormous. The classic story of the extermination of the endemic Passenger Pigeon that once darkened the grassland skies, so huge were its flocks, is only the best known of many.

From being probably the world's most abundant land bird and possibly the most numerous of all the world's birds at the start of the nineteenth century, the Passenger Pigeon was wiped out by the beginning of the twentieth century. Hunters killed the birds up to a hundred per shot and a

A Bald Eagle has caught a sizeable salmon at Kachemak Bay, Alaska. This state, together with Canada's neighboring British Columbia, is now the major world stronghold of a species that was widespread throughout the continent prior to European colonization.

This is the only way that the Passenger Pigeon can now be photographed – stuffed. Though early pioneer photographers did manage to take pictures of Passenger Pigeons piled high on carts after being shot, no one (as far as is known) managed to photograph this bird alive and well in the wild. Once arguably the world's most abundant bird, it took only 20–30 years to wipe it out – an act of wanton folly that even today has its parallels elsewhere in the world.

wagonload a day was not an exceptional bag. Over a thousand were taken with a single throw of a trap net. Many found their way to the markets, but some were shot for some peculiar thrill and left to rot where they dropped. By the time it was realized what was happening to these birds it was already too late, for with a highly gregarious species like the Passenger Pigeon, survival is only possible in numbers. So even when the greatly diminished population became too scattered to be worth hunting, the decline continued. The last breeding record was in 1886 and the last sighting three years later. The last bird was a female named Martha, who died in the Cincinnati Zoo in 1914. The speed of the destruction is perhaps the most surprising factor in this saga, for as late as 1878 millions of birds were still passing into the hands of dealers. Yet only 36 years later the Passenger Pigeon was extinct.

A quite unrelated bird, the Eskimo Curlew, was known as the Prairie Pigeon because of its abundance and suitability for the pot. It too occurred in extraordinary numbers as it migrated eastwards from Alaska and adjacent Canada to the east coast every autumn. There it gorged itself on berries, taking on fuel in the form of fat, preparatory to its migration along the eastern seaboard of the United States, then out over the Caribbean and Atlantic to make a landfall on its wintering grounds among the pampas of Argentina. Inevitably the "Curlew season" in New England became

popular with hunters and it was also hunted during its spring migration through the central prairies of the United States. As a result the bird was soon virtually wiped out and for some years was regarded as extinct. However, it was seen again in 1959 along the Texas coast and has since been sighted at a few other locations.

While the Passenger Pigeon along with the other pigeons and doves is generally regarded as of Old World origin, it is hard to say exactly where the Eskimo Curlew originated. That it is certainly a Holarctic bird is evident, but whether it is New or Old World remains a mystery. The same applies to most of the sandpipers and other waders that breed in the north. There is little doubt, however, that the cranes that inhabit the Nearctic are of Palearctic origin. Two species exist in North America, but while the small Sandhill Crane is still relatively numerous, the large, white Whooping Crane has been all but exterminated.

Prior to settlement by Europeans the Whooping Crane was a widespread

Confined to the New World, though not to the Nearctic region, the Brown Pelican is not only unique in being the only member of its family to perform an aerial dive in its search for food, but also in being a predominantly marine species. This adult, backed by an immature bird, is, in fact, taking off by performing a two-footed jump over the water's surface at the Sea of Cortez, Mexico.

A flight of Sandhill Cranes against an impressive backdrop. The sort of romantic view of wildlife that encourages a sense of optimism that is all too often misplaced. Though still widespread and numerous birds, the present population of cranes is but a fraction of what it was a couple of hundred years ago.

Opposite: A Great White Egret, resplendent in its famous nuptial plumes, with a recently hatched chick. Though reduced in numbers by the plume hunters of last century, this magnificent bird has made a significant recovery and has recolonized many of its previous haunts in recent years.

breeding bird north and south of the Canada–US border in the prairies. The vast marshes that were once so widespread here were easily drained in the nineteenth century to provide rich and fertile land. With drainage the Cranes disappeared. Whoopers had always migrated the length of the continent to spend the winter along the Gulf Coast and, inevitably, their regular movements attracted the attention of shooters. By 1938 the species was down to 14 birds and the subject of rigorous protection. It was seen along the central flyway, wintered at the Aransas National Wildlife Refuge on the Texas coast, and bred at unknown marshes. Only in 1955 were the breeding grounds discovered among the wild muskeg regions of Wood Buffalo National Park in Canada. Thereafter, the migrations of these few birds were followed by light planes and their stopovers guarded from illegal hunting. Such an expensive protection scheme is unique, and gradually the birds responded. By 1967 there were 43 birds and by 1971 there were no less than 59. The protection continues, and is now paralleled by a captive breeding and reintroduction program. Such programs are not without their problems and only time will tell whether the Whooping Crane will survive in the long-term. However, in the short-term the news is very heartening with a total of 133 Whoopers now wintering on the Texas coast.

So, in a native avifauna that includes the wrens, mockingbirds, grouse and others, we find that many species, or at least genera, have colonized the continent from the Palearctic. North America has also been colonized from

The Ferruginous Pygmy Owl is widely distributed, but only in that part of the Nearctic region south of the Mexican border. It penetrates the United States only in southernmost Texas and Nevada, where North American listers must seek it out.

the Neotropical region to the south via the narrow land bridge of Panama. As a result, the largest state lists in North America are claimed by the large southern states, California and Texas. At the same time, although no state can claim a larger list than 600 species, the avifauna of Mexico alone is about 1,000 species. Therefore, the "pure" North American birder must, of necessity, visit the Mexican border region to see birds that only just cross the famous Rio Grande. Here one can find species such as Least Grebe, Hook-billed Kite, White-tailed Hawk, Plain Chachalaca, Red-billed Pigeon, White-tipped Dove, Ferruginous Pygmy Owl, Pauraque, Great Kiskadee, Altamira Oriole and others, including a good collection of humming-birds. Many of these birds may be both widespread and common south of the Mexican border, and form an integral part of the

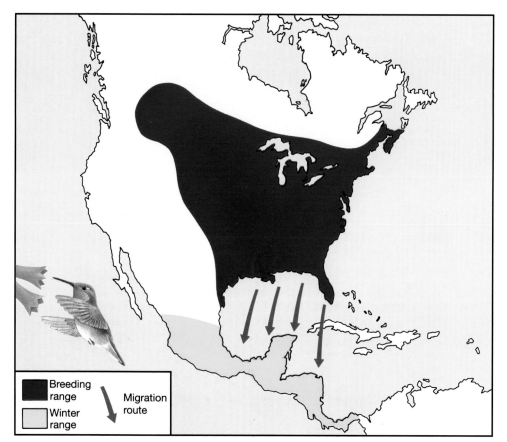

The migration of the Rufous Hummingbird across the Caribbean revolutionized our understanding of migration. It had previously been thought that such tiny birds were incapable of flying such long distances and had to follow either a land or an island-hopping route.

Breeding range
Winter range
Migration route

Nearctic fauna by occurring north of the boundary with the Neotropical zone. They are also useful indicators of birds spreading northwards from the south to colonize the Nearctic.

Of all such birds none are more spectacular than the hummingbirds. Only 13 of the 319 or so species in the world are regular in America north of the Mexican border and some reach no further than southernmost Arizona or Nevada. Eight species can truly be said to be North American and most of these are found in the western United States. Only the Ruby-throated Hummingbird has colonized east of the Mississippi. The Black-chinned is the western equivalent of the Ruby-throated and that bird is replaced in the Rockies by the Calliope and the Broad-tailed Hummingbirds. Most of the other hummers are more or less Pacific coast birds.

The Ruby-throated Hummingbird has been responsible for a major rethink of our theories on bird migration. At less than four inches in overall length and weighing well under half an ounce, this little miracle regularly flies nonstop across the Gulf of Mexico, a distance of some 600 miles. Physiologists stated categorically that this was quite impossible, and it was only by proving that these tiny birds did indeed fly nonstop that scientists were forced to re-examine their theories and adjust them to suit the facts. Now we know for certain that tiny birds are capable of the most extraordinary nonstop migrations over seas, deserts and other inhospitable areas.

Though forming one of the world's most successful bird families, only a few humming-birds have managed to spread northwards from their Neotropical homeland to North America. Two of the most successful are the Broad-tailed Hummingbird of the western mountains (above) and the Ruby-throated Hummingbird that occupies the east (below). Both are summer visitors that make long distance migrations to and from winter quarters.

Although the Ruby-throated Hummingbird extends northwards into the boreal zone of Canada, the Rufous Hummingbird of the Pacific coast ranges as far north as Alaska and breeds no further south than northern California. It is the world's most northerly hummer and winters exclusively in Mexico. On a size basis it must be one of the greatest of all long-distance migrants. Also confined to the west is Anna's Hummingbird which is unique in being largely resident throughout its range. Extending as far north as Vancouver this tiny mite finds food enough to survive even in these high latitudes. Hummingbirds are so small that they must feed virtually continuously on pure nectar to enable them to survive. Many can only live through the hours of darkness by switching all their life-support systems down to a tick-over and passing into a state of torpor. This semihibernation for the night has been found in very few other bird groups and is a unique adaption to size. Hummingbirds are unique in many different ways, but we shall examine them in greater detail where they truly belong, in the Neotropical region.

In contrast to the paucity of hummingbirds, the Nearctic is well served with woodpeckers. There are, in fact, no less than 20 species found north of the Mexican border and many are both widespread and numerous. All are

A female Acorn Woodpecker visits its store of winter food, where the quantity of holes, as well as the mixture of aged and fresh nuts, testifies to extended usage by this particular species. Some trees have been used for many years, though it is not known whether different birds are involved.

A male Gila Woodpecker at its nest hole in a giant saguaro cactus. Famous for providing nest holes that are subsequently occupied by Elf Owls, these woodpeckers penetrate the United States only in southern Arizona and adjacent California, a range that, perhaps not surprisingly, is exactly mirrored by that of the diminutive owl.

adept tree climbers, though some feed mainly on the ground, while others have evolved peculiar feeding methods of their own. They also vary in size from the diminutive Downy to the massive Ivory-billed, which is just a shade under 20 inches in length. As well as being the largest, the Ivory-billed is also the rarest of the woodpeckers and is now generally regarded as extinct in North America, though some may still survive among the forests of eastern Cuba. The extinction of the Ivory-billed is due almost entirely to loss of habitat, to the changes that man has wrought on the flooded forests of the southern states where it once lived. Felling for timber and drainage have both taken their toll of a bird that depends entirely on a diet of wood-boring beetles found only in dead and dying forest giants. The survival of the similar, but slightly smaller, Pileated Woodpecker in a wide variety of woodland types is doubtless due to its preferred diet of ants that can be found among fallen trees and dead stumps.

A highly specialized group of woodpeckers are the sapsuckers. There are three species that, between them, cover most of the woodland in North

America. All bore rows of holes in living trees that then act as feeding stations. The sapsuckers return to holes created earlier to feed on the sap and on the insects that have been attracted to feed in them. Another strange member of the family is the Acorn Woodpecker that has evolved the jay-like habit of storing food during periods of plenty to help it through the lean times of winter. While jays bury acorns, this woodpecker bores holes in trees and wedges an acorn in each. Sometimes a single tree may be riddled with holes to form a massive larder, or food store. In general, the practice does little harm, if any, to the trees it chooses, though the species is decidedly unpopular with the power and telephone companies, whose poles it finds perfect for its needs.

Woodpeckers may have evolved in different ways to exploit different foods and use different feeding techniques, but all are equipped with strong, chisel-like bills and sharp climbing-iron feet for their basic wood-pecking existence. All of the Nearctic species excavate holes in trees to act as nest sites and their need for a fresh hole each year creates a surplus of holes that are then adopted by other hole-nesting species. Some species, in fact, are aggressive enough to usurp the rightful owner before a newly excavated hole has been used. The introduced Starling is a case in point, often evicting a Red-bellied or Red-headed Woodpecker. This "foreign" species, along with the similarly "foreign" House Sparrow, has also been largely blamed for the serious decline of the Eastern Bluebird, with which it competes for old woodpecker-hole nest sites. One highly specialized woodpecker is the Gila Woodpecker of the extreme southwest United States and Mexico. Inhabiting desert where trees are decidedly lacking, it excavates nest holes in the tree-like giant saguaro cactus. This, in turn, has enabled the diminutive Elf Owl to move into Gila holes once they have been used and in this way compensate for the lack of natural tree holes.

A group of Neotropical origin that has colonized northwards is the tyrant flycatchers. No less than 36 species of these small to medium-sized perching birds can be found north of the Mexican border, though this is but a small part of a family that numbers over 360 species, the rest of which are found only to the south. The group includes such familiar species as the Eastern Kingbird, Eastern and Western Wood-Pewees and Eastern Phoebe, as well as the flycatchers of the genus *Empidonax*, which are among the most difficult of all the world's birds in terms of the identification problems they pose. A glance at a good North American field guide shows a number of species that look the same. A similar glance at a South American guide, multiplies the problem to almost insurmountable levels (certainly beyond the range of the visiting or inexperienced birder).

The Nearctic does not have one of the world's largest avifaunas, but it does have some really spectacular species, particularly with regard to numbers. In terms of endemism it is poorly served, but the influence that it enjoys from south and west adds greatly to its variety. Moreover, its position as one of the world's two great northern landmasses, providing a summer home for millions of individuals that spend the winter elsewhere, makes the continent a most exciting place to watch birds.

CHAPTER FIVE
AFRICA
THE ETHIOPIAN REGION

Africa is a well-defined continent, with a land bridge to the Palearctic only in the Sinai region. It is therefore reasonable to assume that the Ethiopian zoogeographical region coincides exactly with the continent. In fact, the whole of North Africa forms part of the Palearctic region and the boundary lies somewhere across the central Sahara around latitude 18°–20°N. It is quite clear that North Africa shows more affinities with the Mediterranean than the Sahel, and we should think of the Sahara as being 'a sea of desert', and just as much a barrier to plants and animals as an ocean.

With a sea of desert to the north and seas of water in every other direction it would seem easy to define the Ethiopian region. Yet even where seas exist, definition is the subject of conjecture and debate. At one time the southwestern tip of the Arabian peninsula, roughly the area of the contemporary state of Yemen, was regarded as falling within the present life zone or zoogeographical region. More recently this area has been placed within the Oriental region, which covers most of the rest of Arabia.

Ross's Turaco, one of 22 species that comprise the family Musophagidae which is totally confined to the Ethiopian region. While most are forest birds, some species have become savannah birds, though remaining essentially arboreal in habit.

Opposite: *Arguably the world's most spectacular crane, the diminutive and aptly named Crowned Crane is found in good numbers over many parts of Africa. Recently this species has been split from the South African Crowned Crane, though the validity of such a division is dubious.*

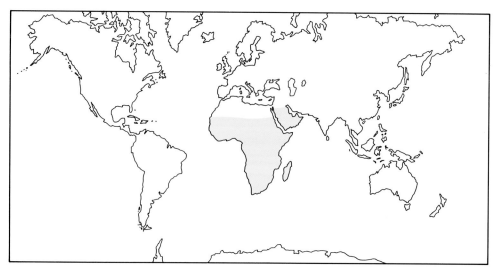

The rounded wings, neatly edged with black, of the Sacred Ibis create a dramatic impact that may have been partly responsible for the esteem in which this bird was held by the Ancient Egyptians to whom it was, indeed, sacred. Frequently mummified and buried in clay pots along with the dead, recent excavations have revealed that many "ibis pots" contained the remains of all sorts of other birds passed off as ibises by hustlers 6,000 years ago.

To the west lies the South Atlantic, to the south the seas that separate South Africa from the distant continent of Antarctica, and to the east lies the Indian Ocean. Yet in the latter direction lies the huge island of Madagascar which, for reasons that will be explained, is often treated by many authorities as a separate region, the Malagasy region, and by all authorities as at least a clear subregion. Mainly on the basis of convenience we shall treat Madagascar as a part of the Ethiopian region, though it certainly merits a separate subchapter. Similarly, the island groups off the coast of East Africa, the Mascarenes, Comoros and Seychelles are all treated as belonging to the Ethiopian region despite obvious differences.

When the Oriental continental plate was detached from Gondwanaland it left behind debris which now forms the world's only granitic oceanic islands. Much, much later, over 100 million years later to be accurate, Madagascar broke away from Africa. Over the past 65 million years the island has developed a remarkable bird fauna that differs considerably from mainland Africa.

Altogether, as defined above, the Ethiopian region holds about 1,600 different bird species, making it the second richest region, after the Neotropical, in the world. Within a single country, Kenya, it is possible for well-guided bird tourists to see over 500 species in a two-week tour and the best record is 700 in three weeks. Before we rush off to book our places on the next birding tour to Africa, it must be remembered that Kenya lies on

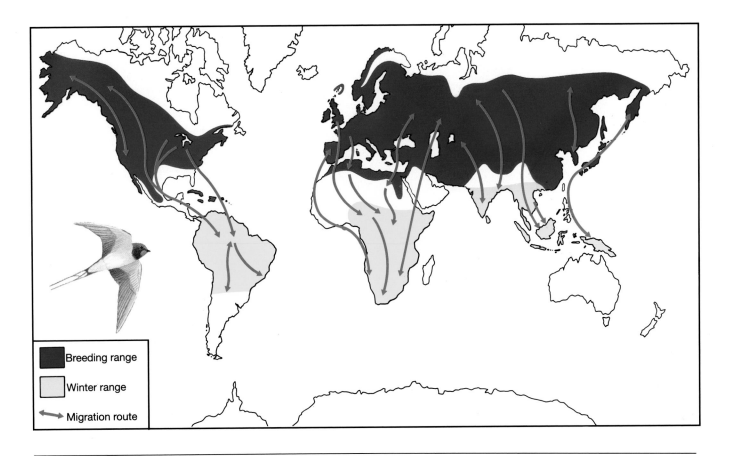

Breeding range

Winter range

→ Migration route

the equator, with its lack of seasonality enabling species to evolve over singularly short distances, and that it varies greatly in altitude from snow-topped Mount Kenya to the baking coasts of the Indian Ocean.

During the last ice age Africa exhibited mountainous conditions from South Africa northwards in a continuous belt to Ethiopia, with a western extension to the highlands of the Cameroons. Across this region montane species were able to move without encountering inhospitable forest or savannah. Today such conditions are found in isolated areas, thereby encouraging the development of distinct species. At the same time, lowland birds were confined to the coasts, to the Congo and to West Africa, likewise encouraging speciation. In this manner, an ice age that never brought ice to Africa had a profound effect on the development of the continent's avifauna. At this time the northern part of the region may have seen a fluctuation in the extent of the Sahara up to 680 miles wide. These changes in climate are largely responsible for the fact that, of the 1,600 Ethiopian species, 250 are found only in the tropical rainforests, 200 are found only above 500 feet, and 100 are to be seen only on Madagascar.

Though Africa is so rich in bird species, many of which are found nowhere else on earth, its close links with other regions are shown by a relative lack of endemic families. Although the actual species may be distinct, other members of the family can be found elsewhere. Even the few endemic African families consist of only one or two species. Mostly the

The Common or Barn Swallow is widespread throughout the Northern Hemisphere and winters in all of the southern continents except Australia. Some individuals make journeys that are among the longest undertaken by any land bird.

The nest of the Hammerkop (right) is one of the largest structures built by any bird. The nest chamber is, however, quite small and approached via a small hole in the base. The theory that a would-be predator is daunted by such a mass may well be true, for nests like this one in the Serengeti, Tanzania are the result of several years cumulative work. The Hammerkop itself (opposite) is a relatively small resident of Africa south of the Sahara and forms an endemic and monotypic Ethiopian family.

region shows its closest links with the Oriental region, though the boundary with the Palearctic is also influential. So the Ethiopian zone's richness in species is largely due to its latitude, climate and landform.

The Ethiopian region is also home to huge numbers of Palearctic migrants outside their breeding season. Though the majority come from Europe and Russia east to the Urals, some come from as far away as eastern Asia, following great loop migrations to return to their area of origin. Ruff, for example, may come from virtually the Bering Straits, while Northern Wheatears originate from the same area, as well as from as far west as Greenland.

Just how many birds fly the Mediterranean and Sahara in a nonstop 40-hour flight is not known, but we are certainly talking about many millions. The late Reg Moreau, a man not noted for exaggeration, made a tentative estimate of 5,000 million land birds making this flight every autumn.

If such a figure seems astronomical it should be borne in mind that the

British Isles, for example, supports a breeding population of three million pairs of Willow Warblers. Double that figure to obtain the number of birds (that is, two members to a pair), add in the production of young (which always outnumber adults at, say, three young per pair) and one reaches a total of 15 million of these warblers alone departing from Britain each autumn. A glance at the range of the Willow Warbler should be sufficient to tell anyone that this total for a small island should be multiplied by a very large factor to calculate the total European population of autumn Willow Warblers. Take into account the fact that this bird breeds right across Siberia almost to the Bering Straits and that even these birds travel to Africa, then the figures do start to become astronomical. Incidentally, Willow Warblers that breed at the far eastern end of their range must fly a minimum of 7,450 miles to reach winter quarters in sub-Saharan Africa.

The migration of birds from the Palearctic to the Ethiopian region is the greatest movement of living creatures in the world. It is not surprising, therefore, that so many ornithologists find the phenomenon fascinating or that it is considered an essential part of the African avifauna.

In terms of Ethiopian endemic families it is a simple matter to list the Hammerkop, Shoebill, Secretary Bird, the guinea fowl and the wood-hoopoes, all of which are found nowhere else in the world. Though the first three are monotypic they are still large and distinctive species, well-known to anyone with even the most passing interest in birds.

The Hammerkop, or Hammerhead Stork as it was formerly and still is

Like the Hammerkop and Shoebill, the Secretarybird, seen here at its treetop nest in Etosha, Namibia forms an endemic and monotypic family in the Ethiopian region. Though related to the birds of prey, it spends its days tramping the grassland savannahs in search of animate food including reptiles and snakes.

occasionally known, is a medium-sized stork-like bird with a peculiar tuft of feathers at the rear of the crown and a heavy bill at the front. The effect is to produce a hammer-shaped head. Its nest, however, is the most obvious outward sign of its uniqueness. This consists of a huge mound of plant debris, sited in the major fork of a tree, that is entered via a hole at the bottom. The work involved in creating such a bulky nest is enormous, but it does make a relatively predator-free home for eggs and young.

The Shoebill, or Shoe-billed Stork, is similarly the only member of its family but is far more restricted in its occurrences than the widespread Hammerkop. Found among the extensive marshes of papyrus that extend from the Sudan to Zambia, it is nevertheless a highly localized species that has relatively few known breeding areas. As a result, little was known of its life-style or of its breeding routines until *Survival* camerawoman Cindy Buxton filmed the Shoebill for television at one of its remote haunts. Seeing the rushes, as unedited film is called, the late Leslie Brown realized that the raw footage was the most detailed écord of the life of the Shoebill and wrote a definitive account for one of the learned ornithological journals on the basis of Cindy's film. The huge bill, the most obvious feature of the Shoebill, is used as a sort of shovel that the bird virtually throws at drying-out mud to catch the large catfish that survive practically without water. In fact, it is almost like diving headfirst into solid ground, a unique method of catching a unique prey under very special conditions.

The Secretary Bird may occupy a completely different habitat, but is

By filming the Shoebill feeding and at its nest, Survival camerawoman Cindy Buxton revealed facts about this unique and shy bird that were completely new to science. Here an adult dribbles water over its chick to cool it in the heat of the tropical sun.

unique in another way. Distantly related to the birds of prey, this is one of Africa's great walkers. In fact, the Secretary Bird spends most of its day walking the great savannah plains of Africa. Like a bird of prey, it constructs its nest on top of a flat acacia, and hunts for large living food items, such as lizards and rodents. But, unlike a bird of prey, it hunts on the ground. As a result, it has long, unfeathered legs like a stork. It also has the reputation of being an accomplished snake-catcher and it certainly behaves in spectacular fashion when it does encounter a snake. Using its long legs and grasping talons, it beats and throws the snake around, avoiding danger by jumping in the air in an erratic dance. It is named from the quill-like feathers that extend from the hind crown, which resemble the quill pens used by the bookkeepers of a bygone age.

The Vulturine Guineafowl, named after its vulture-like bare head and neck, is one of eight species of guineafowl, forming a family found only in the Ethiopian region. These are medium-sized, grouse-like birds that scratch a living from the ground. Their equivalents in other regions are the pheasants in Asia and turkeys in America.

The six species of mousebird are all placed in the single genus Colius *and are clearly closely related. Indeed, the two species* C. macrourus *and* C. indicus *enjoy ranges that meet, but which do not overlap. They are generally regarded as forming a super-species, though they are remarkably similar and could easily be thought of as being members of a single species. Although there is more overlap between the remaining four species, nevertheless there are many areas where only a single species exists. Only* C. leucocephalus *lacks a range that is not, in part at least, free of other mousebirds.*

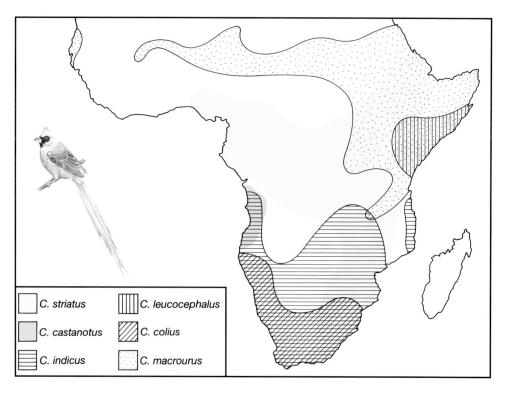

☐ C. striatus	▥ C. leucocephalus	
▦ C. castanotus	▨ C. colius	
▤ C. indicus	⋯ C. macrourus	

The guinea fowl are also ground-dwelling birds that are confined to Africa. Though often thought of as birds of open ground, some are decidedly woodland-oriented in habitat. There are eight separate species, all marked by an upright stance, long neck, short tail and strong, powerful feet and legs. They bear some resemblance to the chicken of Asia and live by scratching food from the ground in much the same manner. Gregarious birds and found in flocks, they are much prized for their culinary qualities. In fact, one species, the Helmeted Guinea Fowl, has been widely domesticated and is bred for food in many parts of the world. This species is naturally at home among the bush country of the savannahs, being replaced in more arid country by the Vulturine Guinea Fowl and in woodland by the Crested Guinea Fowl.

Turning to the smaller birds, the Ethiopian region boasts only three endemic families – the mousebirds, wood-hoopoes and the turacos. There are six species of mousebird, which all belong to the same family and also form the only African endemic order. The fact that the six also belong to the same genus makes them a decidedly peculiar group of birds. This means that we have six species of closely related birds that have no close living relatives. Between them they cover most of sub-Saharan Africa, save deserts and tropical rainforests.

All mousebirds are buffy-brown, marked by short bills, crests, rounded wings and long tails, and differing mainly in head color. They are

Another African family found nowhere else in the world is the mousebirds or colies. The Red-faced Mousebird, seen here at its nest in a protective spiny shrub, is found in West Africa from Zaire south to the Cape. Its long tail is held over its head, in a somewhat uncomfortable position, while incubating.

This Green wood-hopoe belongs to a totally Ethiopian family consisting of eight separate species. All these wood-hoopoes are placed in the single genus Phoeniculus *indicating the close relationship between them and the fact that all originated from a single ancestor. Though usually placed near the Hoopoes in the systematic order, the similarity between the two families may be more a matter of habitat, behavior and plumage pattern than actual physical relationship.*

gregarious birds that seem ill-equipped for both flying and perching. A flight of more than 100 metres is somewhat exceptional and is usually undertaken by each member of the flock in turn, as if waiting for one to arrive safely before the next takes off. When perched the birds usually grip separate twigs with their legs wide apart and the weight of their body hung between them, rather like an avian version of "doing the splits". They feed mainly on buds and fruit and are highly sedentary. As a result, there is considerable geographical variation within the more widespread species. The most obvious of these is the Speckled Mousebird, of which no less than 18 distinct subspecies have been recognized over a range that extends from Nigeria to Ethiopia and southwards to the Cape of Good Hope.

Two species between them cover an even larger range. The Blue-naped Mousebird extends from The Gambia to Tanzania, while the Red-faced Mousebird ranges southwards from Tanzania to the Cape. Though they have been found within a relatively short distance of one another where their ranges meet, there is no overlap and the two are thus regarded as a superspecies – which means that the two species have only separated comparatively recently and their ranges remain separate. Later, perhaps, they will overlap and still not interbreed. Zoologists use two distinct terms

The attractive Hartlaub's Turaco is one of 22 members of this endemic African family. While some five species are largely gray savannah birds, the rest, including Hartlaub's Turaco, are forest birds that run with great agility along the larger branches of fruiting trees and fly on bold, rust-red wings among the canopy.

for such species: those that do not overlap are referred to as "allopatric"; those that do overlap are called "sympatric". In the Coliidae we have two sympatric species and four allopatric species. But the latter are still so closely related that they are often referred to as a species group.

Wood-hoopoes, of which there are eight species, form another group of closely related species by virtue of all belonging to a single genus. Unlike the mousebirds, however, members of this family clearly have strong affinities with the Hoopoe and the hornbills. Like those birds they are highly arboreal, though individual species vary considerably in the type of landform they inhabit. They are basically dark, almost black, birds, washed with various shades of metallic gloss, and they have long graduated tails, long decurved bills, but only weak perching legs and feet with the third and fourth toes partially fused like those of a swift. Though they cling to tree trunks and nest in tree holes, they are unable to excavate their own nests. Between them they range over much of the Ethiopian region, though they are absent from tropical rainforest, deserts and, surprisingly, from much of South Africa. By far the most widespread of the species is the Green Wood-hoopoe, which is so successful that its range overlaps that of most of the other seven species. These are gregarious and generally noisy birds that forage among bushes and trees with acrobatic skill in search of insects, which form the basis of their diet.

The final endemic family of the region, excluding Madagascar, is the turacos. They are large woodland birds, many of which are boldly marked

A Kori Bustard puffs out its neck and raises its tail in display at Namibia's Etosha Pan National Park. Despite its size (it is one of the world's heaviest flying birds) it apparently maintains its numbers well among the plains that it shares with so many predators. Other bustards that have colonized the Palearctic, Oriental and Australian regions have not fared so well and the family as a whole is declining fast.

in shades of brilliant green and with rich chestnut wings that are visible when they fly. They are great climbers, frequently running and jumping along and among the branches of great forest trees. For such a life-style they have strong legs, rounded wings and a long, broad tail, all features making for agile flight and good acceleration. All turacos are fruit eaters and, though they find an easy living in open forested country, there is a considerable level of competition between the various species.

Five species of turaco are gray, rather than green, and have adapted to the open bush country that is so commonly found over large parts of Africa. Members of this group are commonly called "go-away" birds (after their call) and plantain-eaters (after their believed, but incorrect, preference for bananas or plantains). These are large gregarious birds that sit openly atop an acacia or other tree and are among the easiest of all the turacos to see. The group is readily divisible into two superspecies, one of which is spread from West Africa to the Rift Valley, the other from Ethiopia south to the borders of South Africa where, incidentally, it is called the Gray Loerie.

The division of the turacos into superspecies is easily done, for most of the "green" species overlap with only one other species, while their nearest relatives occupy a separate range. Thus two quite distinct green superspecies can be recognized, as can another superspecies among the two violet-colored species. This clear-cut division between the range of one species and the next may be useful to the travelling bird-watcher, but it also indicates the potential for competition between species.

One turaco does, however, stand out from all other groupings in this family – the Great Blue Turaco that ranges across the forested areas of Africa from Senegal to the Congo basin and Lake Victoria. This is a huge, pale-blue bird that is mostly seen flying in small groups from one forest giant to another. Though it doubtless consumes larger fruit than the smaller species alongside which it breeds, it seems unlikely that this factor alone would preclude competition with them.

Aside from the endemic families of the Ethiopian region, it is relatively easy to pick out families that originated there and have since spread to other zoogeographical regions. The Ostrich, which has extended into the adjacent Palearctic in the past and will probably do so again, is an atypical example. More representative of these African families that have since colonized elsewhere are the bustards. Of the 22 species in the world, 15 are African and 14 of those are endemic to the Ethiopian region. Not surprisingly bustards have spread outwards from the Ethiopian region to the Palearctic which has three species, more than elsewhere. Yet there are still two species in the Oriental region and one in the Australian. The latter is a member of a small genus, consisting of four species, which together form the most widespread superspecies. The Arabian Bustard is found throughout the Sahel and overlaps in northwestern Kenya with the Kori Bustard which occurs throughout East Africa and again in southern Africa. The superspecies is represented in the Oriental region by the Great Indian Bustard and in Australia by the Australian Bustard. Though there are

These Red-billed Francolins have gathered at a waterhole in the late afternoon at Etosha Pan, Namibia. Like so many other desert-dwelling species, the need to drink poses a daily risk that predators are ever ready to take advantage of. Thus the Francolins arrive stealthily, drink quickly and are away as fast as possible.

111

Though not confined to the Ethiopian region, shrikes, like these Eastern Long-tailed Shrikes, are probably African in origin. These birds are found throughout eastern and southern Africa and are among the most gregarious members of the shrike subfamily, frequently perching comically on thin saplings that bow under their accumulated weight.

plumage differences between these widely spread birds, they share so many common characteristics that their relationship remains obvious. These are among the largest of all flying birds and a large male Kori may weigh up to 40 pounds. This widespread savannah bird of the African plains is frequently used as a handy perch by Carmine Bee-eaters.

Being large, terrestrial and frequenting mainly open ground, bustards are easy to hunt. They are also easily deprived of habitats when man moves in with mechanized agriculture. Most bustards are therefore in decline and some will doubtless soon gain a place in the *Red Data Book* of species in danger of extinction. These birds have had to face a new danger over the past 20 years as oil-rich Middle Eastern sheiks have pursued them with trained falcons. The Houbara Bustard, perhaps the favorite prey, has virtually been wiped out in Arabia and is now pursued in Pakistan, India and wherever else it can be found. The bustards are a splendid family of birds that deserve more attention than they are getting at the moment.

Among a wide range of other birds that find their natural, if not their exclusive, home in the Ethiopian region, the francolins, larks, cisticolas and shrikes are all notable. To the visitor, however, it is more often the sunbirds that gain the attention. The sunbirds are the Old World equivalent of the New World hummingbirds. But while both groups are boldly clothed in metallic irridescent plumage, feed on nectar taken from flowers, and exhibit a high degree of specialization in feeding techniques, the sunbirds do not hover and, thus, do not produce a humming sound with their wings.

Levaillant's Cisticola is one of the more widespread members of a large genus of Old World warblers. Virtually without exception these are small, brown birds of grass and marsh. Some are named after their songs such as Whistling Cisticola and Chattering Cisticola, others after people such as Chubb's and Carruther's Cisticolas. Quite how the Lazy Cisticola was named, it is difficult to say.

Instead, they hang with great agility to probe deep inside a flower to obtain nectar. Most have long, decurved bills, but none are as specialized as some of the single-flower-based species of hummingbirds.

Within Africa some 70 of the world's 120 or more species of sunbird can be found, indicating that this is certainly in origin a family of the Ethiopian region. Yet despite such a wealth of species they are usually divided between only two genera, showing the close relationship that exists between them. The genus *Nectarinia*, for example, contains no less than 50 African sunbird species. Surprisingly, these birds have not really evolved significant ecological differences, several species feeding not only within the same habitat, but also literally together in the same tree. So instead of evolving distinct food or behavior patterns, the sunbirds have developed

Though bearing a strong physical resemblance to the New World hummingbirds and sharing similar food and lifestyle, the Old World sunbirds are positively unrelated. They are, in fact, the most abundant jewels of the African avifauna though the Lesser Double-collared Sunbird (opposite) is much easier to see than the Bronze Sunbird (above). Though the Double-collared is seen feeding hummingbird-like in flight, this is not the normal sunbird feeding technique.

into separate species on the basis of the distinctive plumage of the male. As anyone who has watched these birds in Africa will know, the females are remarkably dull and similar in appearance.

Though members of the same genus, many species have developed long tails and some of the most spectacular plumages of any bird. The Golden-winged and Bronze Sunbirds, for example, are absolute jewels, their colors heightened by their tails flapping as they fly. Most sunbirds are, however, short-tailed, and it has been argued that these are much more successful and therefore more widespread than the larger-tailed species. Many short-tailed birds are green and are separated by bands of color, often on the breast.

Inevitably, many of these birds have been grouped into superspecies (non-overlapping) or species groups (at least partially overlapping). The collared sunbirds, for example, form a clear-cut three-bird superspecies, while the double-collared sunbirds make up a complex species group. In working out such relationships, systematists are guided more by the dull plumage of the females than by the brightness of the males. In addition, the factors of size, bill length and bill curvature are all highly variable and unreliable features on which to attempt to understand relationships.

For the casual visitor to the Ethiopian region only one family of birds attracts as much attention as the sunbirds. The weavers are abundant,

highly visible, noisy, and confident alongside man. They come readily to feeding stations and drinking fountains, breed freely around safari lodges and hotels and would be dismissed as "sparrows" were it not for their bold colors. In fact, really bold colors, often yellow and black, are reserved for the breeding male. At other times these are dull sparrow-like birds that pose real difficulties in identification.

Of the 52 species of the genus *Ploceus*, the typical weavers, no less than 48 are found in Africa. All of the bishops and widow birds are also African, as are the malimbes. Taking the family as a whole, and thus including the parasitic whydahs, only the sparrows and snow finches have managed to spread outwards to other regions.

The most abundant of the weavers are the queleas, which have been divided among four distinct species. The Red-billed Quelea is often called

the "Locustbird" because of its periodic appearances in huge nomadic flocks that devour everything in their path. In fact, the numbers of this species when it erupts from its breeding grounds are as dramatic as any other of the world's birds. Some authorities treat it as the world's most abundant land bird, others as the most abundant of all the world's birds. Like the other weavers it breeds colonially, often over water. Actual numbers are beyond estimation, for after a day raiding crops the Red-billed Queleas will descend on their communal roost in hundreds of thousands, if not millions. Similarly, at its breeding colonies, usually in dry bush country, nests may number up to 400 in a single tree and colonies vary from an acre to several square miles in extent. If numbers can only be estimated, the extent of damage to crops has been accurately measured. In Senegal 24 acres of rice were consumed in 14 days. Elsewhere one flock, a million strong, was calculated to be eating 60 tons of wheat a day. Even plantations of young trees have been destroyed by the sheer weight of roosting birds.

Left and Opposite:
Two photographs that attempt to capture the extraordinary numbers of Red-billed Queleas that often invade agricultural areas of Africa after breeding among dry, lightly populated savannahs. Probably the world's most abundant land bird, the Quelea is a major agricultural pest that has defined all efforts to control its numbers.

Not surprisingly, waging war on this formidable pest has involved a huge range of techniques including setting fire to the birds' roosts with gasoline, dynamiting, and aerial spraying with poison. Despite the success of such methods, which involved killing 124 million nestlings in French West Africa, 5 million adults in Kenya and so on, huge numbers continue to exist and destroy crops in many parts of the continent.

While queleas and most other weavers construct dome-shaped nests suspended from a tree or bush, the remarkable Social Weaver builds a tenement that looks like a haystack in a tree. Here, each bird has its own entrance and nest chamber and as many as a hundred pairs may nest communally in this way. Though each pair is monogamous and rears its own young, all members of the colony take part in constructing and maintaining the "common" parts, for the internal walls are flimsy compared with, say, the thatched roof.

The whydahs are a small group of weavers that are brood parasites of the waxbills, which they often closely resemble. The males are mainly black during the breeding season, with lengthy tail extensions that are especially dramatic in flight. The male Paradise Whydah, for example, has a tail consisting of four central tail feathers that are almost four times as long as its body. In flight, two are held vertically and two horizontally to catch the air in the most extraordinary way. In point of fact these birds fly awkwardly to cope with their unaerodynamic shape and balance.

Although this Social Weaver seems to have the huge nest to itself, it does in fact occupy only a single tenement in a giant block of avian apartments. As many as a hundred pairs may share in constructing the great tree nest, though each pair then has its own nest chamber in which to incubate its eggs and rear its young.

Like the parasitic cuckoos, the whydahs lay eggs that closely mimic the species that they depend on. Even the mouth markings of the young resemble those of the host's chicks. Research has shown that each whydah relies on a particular species of weaver-finch to rear its young. So each species is dependent on the success of its chosen host. If the host population is reduced or exterminated, then the whydah, too, is at risk. However, in captivity, breeding without a host has been proved for some species, raising interesting questions about brood parasitism in general.

Among the most widespread of the weavers is a group of mainly black and yellow species that are easily confusable to the casual visitor. Most are yellow below, black and yellow above and marked by a black mask. The extent of the mask, the color of the eye and the extent of rust at the mask border are the best identification features, but outside the breeding season many species are most difficult to differentiate. A knowledge of where each species occurs is often crucial and always helpful.

Three further groups of Ethiopian species are worthy of more than a passing mention. The two oxpeckers have evolved a symbiotic relationship with the population of large mammals that inhabit the African savannahs. Virtually identical apart from bill color, the Red-billed Oxpecker is essentially eastern, the Yellow-billed more western. However, the two species overlap in huge areas of southern and eastern Africa, the very areas where the largest herds of mammals survive. These birds are closely related to the starlings and are regarded as no more than a subfamily by current systematists. It is perhaps not surprising, therefore, that elsewhere in the

The Shaft-tailed Whydah is one of ten species of highly specialized weavers that have become brood parasites of the waxbills that they closely resemble. Males, like this one, in breeding plumage develop long tails that alter their flight balance and produce attractive and highly visible aerial displays. ·

world starlings are often found perching and feeding on domestic stock.

Also closely associated with the great game herds of Africa, though hardly symbiotically, are the vultures. Soaring effortlessly hour after hour, they are among the most obvious of African birds. These Old World vultures are not related to the vultures of the New World, but are close to those of the Palearctic and Oriental regions. By far the most abundant virtually throughout its range is the African White-backed Vulture, a general-purpose scavenger unrestricted by the specialized tendencies of other larger and smaller species. In densely settled areas, where large game has been destroyed, the smaller Hooded Vulture seems more able to live alongside man than any other species. So, as the African human population explodes there has been a decline in the larger vultures and an increase in the

Relatively few species of parrots inhabit the Ethiopian region, though the Senegal, or Yellow-billed parrot (opposite), is common from The Gambia to Cameroon. The Masked Lovebird (left) of Tanzania is one of a number of species of small, gregarious parrots that cuddle-up close to one another and often preen each other in a "loving" sort of way.

smaller species. Strangely enough, such a shift in population fortunes has not occurred, despite the huge human population, in the Oriental region, where the all-purpose scavenger is the medium-sized Indian White-backed Vulture. Perhaps the plentiful supply of large carcasses – the cow is after all sacred to Hindus – is a more important factor than the presence or absence of large numbers of humans.

No comparative account of the birds of Africa would be complete without mention of the Yellow-throated Longclaw. This yellowish bird inhabits open country with a good covering of grasses and other low vegetation. Several thousand miles away in the similar landscape of North America, lives the Eastern Meadowlark. The two are nearly identical, but completely unrelated. The likeness is totally due to convergent evolution, that is, birds living in the same habitat in the same way coming, over time, to look the same.

The Madagascar Subregion

Despite the evidence, zoogeographers have always shown a remarkable disinclination to treat the island of Madagascar as a zoogeographic region in its own right. This author, though no zoogeographer, continues this well-established fudging of the issue – a situation due in part to the relatively small size of Madagascar, which covers a mere 250,000 square miles, but also to the proximity of the island to Africa and the great Ethiopian region. Yet the fauna of Madagascar is certainly as different from its giant

Lesser Flamingoes gather in tight formation in a "stomping" display that forms an integral part of bringing the whole population to the point of breeding readiness. Such communal displays are crucial to a species in which synchronized breeding is essential for success.

neighbor as that of the Oriental region, and, indeed, it shows a level of endemism higher than either of these larger regions. Thus, at worst Malagasy is a subregion, and based solely on the evidence there is an outstanding case for regarding it as a full region in its own right.

Madagascar lies south of the equator between latitudes 12° and 25° south, more or less east of Mozambique, which it closely parallels in size. At its closest it is no more than 250 miles from the coast of Africa. The Comoro Islands, which lie between Madagascar and Africa, are, quite reasonably, regarded as forming part of the region (or subregion), but the inclusion of the Mauritius and Seychelles groups is somewhat less justified. Nevertheless, for our purposes Madagascar is a well-defined region by virtue of being surrounded by sea in every direction.

Before the great continents Gondwanaland and Laurasia started to break up 180 million years ago, what is now Madagascar was neatly sandwiched by the present Ethiopian and Oriental regions. The Indian continent then drifted away to crash into Asia some 120 million years later. Whether Madagascar split from Africa at the same time or remained attached is the subject of conjecture. Certainly, the fauna of Madagascar, including its avifauna, is quite distinct from either of the regions that once formed the bread to its sandwich. Outstanding among these are the lemurs, an endemic group of mammals that replaces the monkeys found in Africa and India.

The huge Goliath Heron spreads its wings to lift off from lakeside vegetation. Though a typical heron in habit, it builds its nest among floating papyrus beds rather than in trees, a practice shared with the Purple Heron and Great White Egret.

When it comes to birds Madagascar boasts no less than six endemic families (only three according to some authorities, though all rank them as at least meriting subfamily status). The three upon which all authorities agree are the mesites, asitys and vangas. The disputed three (as full families) are the Cuckoo-Roller, ground-rollers and Coral-billed Nuthatch. Even here confusion is rife, for some consider "Nuthatch" as a complete misnomer for the Malagasy bird, regarding it as having no affinities to the regular nuthatches, but simply as an aberrant member of the vangas. Nonetheless, the fact remains that the Madagascan avifauna shows a remarkable level of endemism at family level.

At specific level the uniqueness of Madagascar is even more apparent for no less than 65 percent of its breeding birds are found nowhere else on earth – of 182 breeding species, 118 are endemic. When it comes to the passerines the level of endemism rises to a staggering 95 percent, a proportion only found elsewhere on the smallest and most isolated oceanic islands.

The fish-rich lakes of the Ethiopian region provide a home for healthy populations of the African Fish Eagle, a bird that has close relatives in every other part of the world. At the best fishing grounds the most powerful birds occupy territories along the shoreline, while the less powerful nest away from the banks and are forced to hunt in the middle of the lake where perches are few or non-existent. This immature bird is about to catch a fish at Kenya's Lake Baringo.

Such proportions of endemism indicate not only that Madagascar has been geographically isolated for an extremely long time, but also that its avifauna is not the result of recent winged colonization. This conclusion is supported by the fact that the 118 endemic species are highly differentiated and placed in a large number of distinct genera. Thus the average Madagascan genus consists of less than two member species – actually 1.3 species. Madagascan birds exhibit little of the species radiation found in isolated islands such as the Galapagos or Hawaii. It is easy to conclude that these birds have been established here for so long that their relationships, one to another, have become blurred by evolution, but of all Madagascan families only the vangas share any close relationship.

This endemic family, often called vanga–shrikes, consists of 12 distinct

species that are usually placed close to the true shrikes in systematic lists. They are small to medium-sized perching birds with heavy, sometimes hooked, bills and are generally boldly colored. Mostly arboreal in habit, they feed on large insects that they find by scouring the outer branches of trees. Some of the larger species regularly take small tree-dwelling frogs and lizards. They are gregarious and noisy, moving through woodland in small or large groups. While bill shape and size is the most obvious difference between the various species, so little is known about these birds that conclusions regarding their relationships, and thus their evolution, are impossible. For instance, some authorities wish to make the Coral-billed Nuthatch the thirteenth species, and others want to place the Helmetbird in a family of its own.

Of all Madagascan birds none are as thought provoking as the elephant birds. These large and flightless birds all became extinct before historical times, though the legend of the Roc and its relationship with Sinbad the Sailor may indicate that some species survived until the arrival of humans. It is just as likely, however, that the discovery of one of their huge eggs,

The African Jacana, or Lilytrotter, finds a hippopotamus a handy perch among the weed-clogged waters of this nutrient-rich lake. The presence of hippos can often create such conditions by virtue of their huge intake and "output" of vegetable matter.

A Hoopoe feeds its well-grown young at a tree-hole nest site. This is one of the Old World's most successful birds that ranges through three of the world's zoogeographical regions. Inevitably it has formed a number of distinct subspecies – some authorities treat them as full species – including Upupa epops marginata *in Madagascar.*

preserved in a marsh, led to the growth of such a terrifying legend. The largest elephant bird egg, which measures about 13 × 10 inches, is certainly the world's largest single cell and it is easy to imagine the myth of the gigantic and ferocious bird that laid such a monster.

But, although the Roc is pure legend, the existence of the elephant birds is not. The largest elephant bird, *Aepyornis titan*, was a ten-feet-tall monster of enormous bulk. Some of the moas of New Zealand may have been taller, but none was as bulky as *A. titan* which weighed in at a staggering 1,000 pounds. It is easy to relate such large flightless birds to the present day Ostrich, Emu, cassowary and rhea, as well as to the more recently extinct moas. Nonetheless, the problem of their origins remains. Did they evolve before the continents split apart, or are they the result of colonization by

The Crested Madagascar Coucal is placed in a family or subfamily the Couidae (Couinae) that consists of nine different species, all of which are confined to Madagascar. It is usually placed between the ground cuckoos of Asia and the true coucals of Africa and Asia, though there is still great doubt about its true origins.

Carmine Bee-eaters are among the most colorful and gregarious members of this highly attractive family of birds. Creating dense colonies among the large rivers of southern Africa, they are as much a part of the African scene as any of the great mammals.

The male Madagascan Red Fody in full breeding plumage is a crimson gem marked by a black eye patch. Outside the breeding season it becomes a sparrow-like bird similar to its mate in browns and buffs. Its introduction to Mauritius, the Seychelles and elsewhere has had an adverse effect on local fody species.

flying ancestors? If the former, then we have to revise our ideas about the timing either of the splitting up of the continents or of the evolution of birds.

The idea that the elephant birds may have derived from the African Ostrich, or that these birds may share a common ancestor, posits a closer link with the African avifauna than there is evidence to support. Comparing the birds of the two regions we find not only that Madagascan birds are of different species, but also that the very structure of the island's avifauna is quite different. Thus that spectacular and abundant African group, the weavers and finches, is all but absent in Madagascar, which has only three weavers and a single finch. In contrast, over 30 percent of Madagascan birds are aquatic, but only eight percent of African birds are water dependent. And yet, while it is water birds that would find it easier to colonize from the mainland, in actual fact no less than 30 percent are endemic to Madagascar.

The Blue-cheeked Bee-eater is the only member of its family to be found on Madagascar. This is a widespread bird that ranges from the Middle East to Central Asia and Africa. Birds that breed in Madagascar winter in Africa and may well have colonized from the east, for several authorities regard the Madagascan subspecies as identical to these oriental birds.

It is thus a reasonable supposition, if not exactly clear, that Madagascar has been separated for a sufficiently lengthy period to have evolved its own distinct avifauna showing little connection with the present-day bird life of nearby Africa. Sadly, although the birds of Madagascar have been so little studied, the evidence that remains in the form of living birds has been woefully depleted. Habitat destruction and degradation is as severe here as anywhere in the world, and the remnants of the once extensive forests are no more than widely scattered pockets of isolated habitat. Birds and other animals, including the memorable lemurs, are still extensively hunted either for food or as pets. In this fashion, one of the world's most fascinating faunas is disappearing before scientists have had a full opportunity to study it. That Madagascar is unique is not in question. What is in danger is the evidence that it holds for understanding the way in which life evolved as the continents split apart.

CHAPTER SIX
ASIA
THE ORIENTAL REGION

The Oriental region extends from Pakistan in the west to the Philippines in the east, from China in the north to Java in the south. It is thus bisected, in Indonesia, by the equator, but extends northwards beyond latitude 30°. The boundaries abut the Ethiopian, Palearctic and Australian zoogeographic regions, though in most cases these are reasonably well defined.

To the south the sea boundary is formed by the Indian Ocean and the Bay of Bengal. In the north, the great Himalayan chain together with the Hindu Kush forms a mountainous boundary that, being inhospitable to most life forms, is an effective barrier between this region and the Palearctic. In the east the Pacific forms a marine boundary, and in the west the Arabia Sea is another. If these broadly drawn boundaries seem reasonable enough, they are not without their problems. Indeed, drawing the boundary between the Oriental and Australian regions has proved as difficult as any problem confronting zoogeographers anywhere in the world. Other boundaries, however, are not without their difficulties. In the west the boundary is

The head plumes of the Golden Pheasant are among the most splendid of this often beautifully adorned family, that is widely spread in the forests and hills of the Oriental region.

Opposite: *Ornithologists more familiar with the Ethiopian fauna could be forgiven for identifying these birds as Yellow-billed Storks. In fact they are the Oriental equivalent, the remarkably similar Painted Stork which is equally gregarious both when feeding and nesting.*

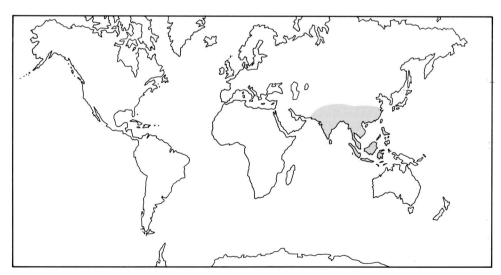

One of the most widespread and successful of the world's birds, the Great White Egret shows its nuptial plumes to perfection as they catch the light while the bird preens. The people of the Oriental region, with their deep-seated religious regard for life, seldom persecute the birds that live happily alongside the human population.

generally drawn along the line of the Indus to its mouth at Karachi, thus bisecting Pakistan. This places the whole of Iran and Afghanistan in the Palearctic. In the northeast the boundary runs along the Hwang Ho river, cutting China roughly in half. Inevitably, such lowland divisions are debatable, though the experts are in remarkable agreement in this particular area.

However, it is when we come to the southeastern boundary with the Australian region that major difficulties arise. Yet rather than think of such difficulties as being tedious and boring, we should regard them as exciting, indeed, as the very stuff of zoogeography.

The first division of the world into zoogeographical regions was proposed by Philip Lutley Sclater, a nineteenth-century ornithologist, in a scientific paper presented at London's Linnaean Society in 1857. Sclater's conclusions, based on studies of birds, were confirmed some 20 years later by Alfred Russel Wallace following studies of other animals. The regions proposed by Sclater are more or less those followed today. It is the work of Wallace in Malaysia, however, that concerns us here. His intensive studies showed that some birds and animals were present on one side of the narrow strait between Bali and Lombok but absent from the other, and vice versa; and the, albeit wider, straits between Borneo and the Celebes produced similar results. This boundary was later referred to as "Wallace's Line", the division between the Oriental and Australian regions.

While Wallace drew his boundary largely on the basis of the presence or absence of species typical of the different faunas, more recent studies here have tended towards a more statistical approach. Of course, it is well known that present-day zoologists, particularly ornithologists, are so blinded by statistics that scientific papers lacking "significance tests" are virtually worthless. Nevertheless, on a statistical basis Max C. Weber drew a line well to the east of Wallace's Line, thus including the islands of the Celebes and Timor in the Oriental region. Such a line is justified by finding points where a balance between Oriental and Australian elements exists. This more modern approach is now widely accepted and the boundary between the two regions follows that proposed by Weber in 1902, although the area between the lines proposed by Wallace and Weber is still referred to as "Wallacea" by many zoogeographers.

From an ornithological viewpoint this area remains of great interest and it is surprising that the opening up of Bali as a modern tourist paradise has not seen masses of ornithologists scuttling off on cheap package holidays to follow in the footsteps of the nineteenth-century greats. In Bali, for example, there are barbets, whereas only 15 miles away in Lombok they are entirely absent. Barbets are widespread in the Oriental region, but totally absent from the Australian. Conversely, the typical Australian honeyeaters occur as far west as Lombok, but are absent from Bali. Many other bird groups show a similar, if incomplete, division between these two islands. In general, it is best to regard the boundary as a transitional zone rather than as a definite geographical line, though traffic across and through the area is mostly one-way, from the Oriental to the Australian.

Though most of the Oriental region lies north of the equator, much of the landmass is decidedly tropical in climate. Rainfall, associated with the various monsoons, is high and forest covers large areas of the lowlands. It is, however, among the most densely populated regions of the world and much native forest has been cleared for agriculture. China and India are the most populated countries in the world and Java is as densely populated as most city suburbs. Birds have had to inevitably adapt to the rapidly increasing human population, or find refuge in the small pockets of natural vegetation that remain. A typical example of wholesale forest clearance can be seen in the Nepalese terai, the area of heavily forested hills that forms the political boundary between that country and India.

Until the 1950s the terai was an impenetrable jungle of tall trees and huge creepers, broken by lowland swamps of dense elephant grass. It was protected in this state by a population of malaria-carrying mosquitoes and was regarded as a death trap by all but a few local tribes who had built up a semi-immunity to this disease. Under a postwar aid program and by the use of an insecticide now largely banned throughout the world, malaria was eradicated from the terai. Within a few years the land-hungry population of the Nepalese hills had settled, cleared and were farming this formerly inhospitable wilderness. Tigers, leopards, the rhinoceros and hosts of birds disappeared from all but a few pockets protected as royal hunting reserves by the Nepalese king. Today, these well-established national parks are no

The White-winged Wood Duck is the Oriental equivalent of the Neotropical Muscovy Duck. Placed in the same genus, these two closely related birds are separated by thousands of miles of the Pacific Ocean yet are deemed to have shared a common ancestor. But which bird colonized which continent?

more than symbols of what, only 40 years ago, was a truly great wilderness.

The effects of human settlement and an ever-growing population have been as dramatic in the Oriental region as they have been in Europe and the United States. But, while relatively few species have been totally exterminated, actual bird populations have undoubtedly declined. Even here, however, there are significant differences between one part of the region and another. China, for example, is noted as a virtually bird-free zone, for not only are birds regarded there as food, but also decrees from a government determined to produce sufficient food for its enormous population stressed the need to kill all small seed-eating birds. The result is an incredible lack of birds over huge areas of the country. India, by contrast, is a land full of birds. This difference is due almost entirely to the Hindu religion, which regards life as sacred. Not only are cows considered sacred and not slaughtered, but in any case a large percentage of the population is vegetarian. Thus birds, as well as other wildlife, live happily alongside man, sometimes in an astonishing fashion. Kites and vultures can be seen in huge numbers over many Indian cities, as well as around villages. They are often quite fearless and will gather at the site of carrion with little regard for people around.

The Oriental region is thus particularly well off for birds of prey and even, as we shall see, acts as a secure wintering ground for these birds from further north. Among the plains of India the large King Vulture is the

equivalent of the Lappet-faced Vulture of Africa. The White-backed Vulture of the Orient has its counterpart bearing the same name in the Ethiopian region. The Long-necked Vulture can be compared with Ruppell's Vulture, while the Egyptian Vulture occurs in both regions, as does the Lammergeier which finds some of its world strongholds among the mountains of Ethiopia, South Africa and the Himalayas. Strangely enough two Palearctic vultures, the Griffon and Black, also range as far as India, making that country the world's vulture headquarters both in species and, doubtless, numbers as well.

If one is privileged, or unfortunate, enough (depending on the strength of one's stomach) to come across a gathering of vultures around a recently dead cow, it is quite amazing how little time is required to clean up the mess. Ornithological textbooks tell us that each of the various species is equipped to deal with certain parts of the carcass – the huge-billed King Vulture to tear open the carcass, the delicately billed Egyptian to pick at the bones and so on. Yet confronted with reality, the vultures seem to forget the rules and pile in helter-skelter. It takes only a few seething minutes until all that is left is skin and bone. India is not the world's tidiest country and it is as well that someone takes over the clear-up operation so effectively.

Indian White-backed and Griffon Vultures gather at a carcass at the Manas Wildlife Reserve, Assam. The fact that Hindus regard the cow as sacred and never kill it, provides a regular and abundant supply of carrion in the form of cattle that die of old age and disease. As a result the population of vultures in India is as healthy as any in the world and a lot healthier than most.

The expanded wattles of Bulwer's Pheasant (right) are backed by a puffing-up of the breast and neck feathers to create a dramatic nuptial display. Confined to the rainforests of Borneo, this species is seldom seen and only rarely kept in captivity.

Opposite: *Sunbirds are far less abundant in the Oriental region, both in terms of species and numbers, than in the Ethiopian region. Most widespread of the Asiatic species is the Purple Sunbird (top), whereas the beautiful Fire-tailed Sunbird (bottom) is found only among the forests of the Himalayas.*

Even in India, however, the strain on the land by virtue of an ever-expanding human population means that natural forest and marsh have largely been destroyed. A network of national parks and nature reserves covers only a tiny percentage of the land, with each isolated from the next, often by hundreds of miles. For highly mobile birds this may pose no great problem, but for large mammals the problems are insurmountable. Tigers unable to establish territories within their natal reserve, for example, are forced into marginal habitats occupied by people, with disastrous effects. Elephants moving through rice fields may not have such dramatic results, but they are equally unwelcome. Man and wildlife can coexist, but it is a precarious existence and one that, in the long term, the animals are bound to lose.

Two high altitude species that, in their different ways, have learned to cope with the thin air of the high Himalayas. The Blood Pleasant (right) lives at over 14,000 feet near the tree line and regularly visits human habitation, virtually unheard of in a wild pheasant. The Bar-headed Goose (opposite) breeds on the Tibetan plateau at similar altitude, but then flies over the highest Himalayan peaks to winter among the plains of northern India. It is thus the world's highest-flying bird.

If we have dwelt overlong on the problems of birds and man it is because so much of the Oriental region is very densely populated. In general, farming is intensive, small-scale and self-sufficient. This means that huge areas of the region are given over to agriculture, with fields extending from the rich lowlands to terraces high into the hills.

The region does, however, have many other habitats and, despite what has been described, some really extensive forests, particularly in the south on the island of Borneo. These rainforests consist mostly of the large hardwoods that are in such high demand commercially, for softwoods are confined to the highest Himalayan mountain slopes. Watching birds in such forests is similar to other parts of the world – total absences for long periods followed by short periods of ferocious activity as a mixed feeding party, consisting of a hundred or more individuals of twenty or more species, passes through. Among them may be the leafbirds that are the only purely endemic family of the region.

Agricultural and bush country holds birds that, though not endemic, are shared only with the Ethiopian region. They include the weavers, sunbirds and honeyguides, but all are outnumbered by their African relatives, which indicates a comparatively recent colonization. Equally characteristic of open country are the starlings and finches, often the most abundant birds in many parts of the region.

Probably the most significant of all Oriental bird habitats are those formed by the great mountain chains that extend along much of the region's northern boundary. Reaching a peak of over 26,000 feet in the central Himalayas, these mountains not only remain some of the least spoiled of the world's wildernesses, but also act as an effective barrier to both colonization and migration. Here the pheasant family is particularly well developed and, taking the region as a whole, many genera of these birds are found nowhere else. High in the mountains the Impeyan and Blood Pheasants reach to the tree line and beyond, the latter becoming quite tame around high-altitude Buddhist monasteries such as the famous Thyangboche along the route to Everest.

Yet, as one climbs higher into these mountains, the birds become progressively more Palearctic in form and, sometimes, in species as well. The Grey Tit of the Himalayas is no more than a paler, washed-out form of the Great Tit, while the Nuthatch, Treecreeper and Wallcreeper are exactly the same as their Palearctic counterparts. Jays, thrushes, redstarts, dippers, and accentors, if not the same species, are all clearly related to Palearctic birds. Two of the highest-altitude birds in the world are, perhaps not surprisingly, found among these mountains. The Snow Pigeon is commonly found around the sherpa villages of Nepal up to 16,500 feet, while the Tibetan Snow Finch ranges even higher.

Yet interesting as the Himalayas are in being a sort of ornitho-melting pot between the Palearctic and Oriental regions, it is their influence on migration that so frequently fires the imagination. Mallory, most famous of the unsuccessful climbers to attempt Everest, reported seeing geese flying over that mountain, which is just over 29,000 feet high, and guesstimated their altitude, quite reasonably, at some 30,000 feet. There can be little doubt that these were the Bar-headed Geese that breed at altitude on the Tibetan plateau and spend the winter among the plains of northern India. Other observers, including myself, have seen Steppe Eagles flying southward from this same massif at 15,000 feet, descending from passes that are at least 18,000 feet high. In such cases the lack of oxygen at these heights must pose a problem for birds, and the active behavior, as against passive soaring, of the geese makes their performance even more astonishing in this respect.

In general, bird migration takes place at low altitudes, mostly below 5,000 feet and, for this reason, migrants tend either to avoid mountains or to concentrate at the lowest passes. The problem with the Himalayas is that even the lowest passes lie at considerable altitude, so many birds avoid the mountains by flying around them. There is therefore a tendency for many Palearctic migrants to winter among the Indian plains at either the eastern or western ends. The mountains thus create a sort of shadow of absence among many winter visitors. Even within a particular species there may be a migrational divide, with some populations taking an easterly route and others a westerly one.

An outstanding case of birds avoiding the higher mountains is the migration of the beautiful, but unfortunately rare, Siberian White Crane. The western population of this splendid bird breeds among the marshes of the Ob river in northern central Siberia and winters only at Bharatpur, near Agra in northern India. A direct route between the two locations passes over the high Himalayas, and so the birds make a detour to the west via Afghanistan and Pakistan.

Not all birds, of course, avoid the mountains and although direct evidence is lacking the concentrations of Demoiselle Cranes in March in southern Nepal would seem to indicate a direct flight through the world's highest mountain chain. Interestingly enough, the Siberian Crane is totally absent from Nepal. The vast flocks of Palearctic ducks that winter on the Indian plains also head directly northwards, with thousands of Northern Pintail heading up the Kosi river valley in spring on a direct line towards the Everest region.

The migrations of smaller birds are inevitably more difficult to detect, though Bluethroat, Siberian Rubythroat, Black-throated Thrush and others

Vast numbers of duck leave their Siberian breeding grounds to winter among the plains of India. During the day this flock of Pintail and Shoveler can be found roosting on an artificial lake in Delhi zoo surrounded by the city's teeming human population.

The call of the Common Iora is responsible for its English name. Though brightly colored, it is an inconspicuous forest bird of the Indian plains.

The Tailorbird, here seen stretching to feed its hungry young, is named after its unique ability to sew leaves together to act as a nest cup. It is a common and widespread species in many parts of the region.

are regular passage or winter visitors to the Kathmandu valley of central Nepal. One could thus surmise a trans-Himalayan migration for these and perhaps many other birds.

Although the Oriental region boasts only one endemic family, the leafbirds, they are a widespread and colorful group. They are essentially arboreal in habitat, mainly frequenting forests, but also occurring in belts of trees in agricultural and bush country. Generally with short legs and small feet and mostly boldly colored in shades of green, yellow or blue, colors that merge easily with well-vegetated areas, they have relatively long, often decurved bills and feed largely on fruit and insects, though some species also take nectar. These are normally highly vocal birds that produce quite musical trills.

The family consists of three fairly distinct groups – the ioras, the leafbirds and the fairy bluebirds. The ioras are the smallest and are generally less gregarious and more insect-oriented than the other groups. The Common Iora is widespread in the region and is perhaps best noted for its dramatic display flights in which the bird fluffs up its feathers to form an aerial ball and drops slowly towards the ground uttering a strangely tin-whistle-like call.

The two fairy bluebirds are gregarious fruit-eaters with a particular penchant for wild figs. The Blue-backed Fairy Bluebird is widespread from

The Golden-fronted Leafbird is a member of an Oriental subfamily that includes the ioras. The species is widespread from the Himalayas through Indochina to Sumatra and is an easily overlooked forest-dwelling bird. Sadly, it is often caught for export as a cage bird and large numbers perish as a result.

The displays of male pheasants and their allies are among the most spectacular in the world of birds. Undoubtedly the best known are those of the Blue Peafowl, or Peacock (above opposite). The magnificent train, which is held folded (below opposite) most of the time, is raised and spread to show the feather "eyes" to advantage and then shimmered before the more dully colored females. Despite appearances, the Peacock's train consists of elongated uppertail coverts, the real tail consists of short, stiff feathers that act as a support. Arguably even more spectacular, though certainly less well-known, is the display of the Great Argus Pheasant (above) in which extended wing feathers are turned inside out.

India to Vietnam, whereas the Philippine Fairy Bluebird is confined to those islands. Some authorities treat both as members of the Old World oriole family. Leafbirds proper belong to the genus *Chloropsis* and are predominantly green, marked with yellow and black. The Golden-fronted Leafbird is widespread, frequently kept as a cagebird and exported in some numbers.

Though these birds form the only endemic family of the Oriental region, several other families are particularly well represented and doubtless have their origins in the region. Largest and most spectacular are the pheasants. Some, as we have seen, are mountain birds, but others inhabit dense hill forest. The Common Pheasant enjoys a native range that extends from China right across the Palearctic to Turkey, it is the world's primary sporting bird and has been introduced throughout Europe and North America. Several other pheasants have also been introduced outside their natural ranges including Reeve's, Golden and Lady Amherst's, all of which are found in the wild only in central China. The males of these three species have spectacular plumage and are kept mainly for their decorative rather than sporting characteristics.

Even more spectacular are the six species of peacock pheasant of the genus *Polyplectron*. Males, at rest, are marked in shades of gray, but each feather of wings, tail and uppertail coverts is marked with a boldly colored disk that produces the most dramatic effect when shown to advantage. The

144

display itself is somewhat strange as the male walks around the female, spreading his finery in a sideways or lateral manner that involves turning the spread wings and tail towards the object of his desire. Should the female be interested he may then perform a full-frontal display guaranteed to show the gleaming and colorful disks to full advantage.

Peacock pheasants are very difficult to see in the wild. They frequent the dense tropical jungles of Malaysia and Indonesia, from Burma and Thailand to Borneo and Sumatra, and are very secretive and skulking. Even today, the nests of some species remain undescribed. Rothschild's Peacock Pheasant, for example, is confined to an area of mountain forest in central peninsular Malaysia. This is tough country and nothing is known about its life at all. Similarly, the Palawan Peacock Pheasant is found only among the dense tropical forests of that island. It is a small, skulking and rare member of the genus and is threatened by the wholesale logging of its native jungles.

The Great Argus Pheasant is also a remarkable skulker. Like the peacock pheasants it has an elaborate display that is marked by showing the disks of wings and tail to advantage. It does, however, have enormously extended and broadened central tail feathers that are frequently twisted sideways towards its mate while the wings are brought forward to form a complete circle. Such dramatic posturings are shown to best advantage on a special stage that the bird constructs by removing leaves, twigs and other debris from a small jungle clearing. This stage or arena is jealously guarded against all intruders. As with other well-adorned birds that perform elaborate displays, the male woos a series of females, which then rear their family single-handed. The Great Argus is found from peninsular Malaysia to Sumatra and Borneo where it inhabits forested hillsides.

Of all the pheasant tribe none is more spectacular, or as well-known as the Blue Peafowl, or Peacock to the layman. Despite its loud and penetrating calls, it is still widely kept as a decorative pet. Males have specially extended uppertail coverts that form the distinctive train and may be up to three times the length of the bird itself. In display the shorter and more rigid tail feathers are raised as a support for the fan-shaped train, each feather of which is marked by an eye-like disk. In full display the whole is shimmered in the most dramatic way.

The Blue Peafowl is widespread throughout Pakistan, India and Bangladesh where it inhabits open bush country. Local people use the cast-off uppertail feathers for a variety of decorative purposes, including fans for tourists. After being kept in captivity for many generations various colormorphs have been bred, including pure white birds, the display of which creates a remarkable ghostlike effect.

One of the strangest features of the Oriental region is the remarkable distribution of birds within the Indian subcontinent. While it is a primary purpose of this book to seek to explain the differences and similarities between the various zoogeographical regions, the presence and absence of species within a single region may exhibit patterns that are no less interesting. So, for example, many of the birds of China, Malaya and the eastern Himalayas are completely absent from peninsular India, but occur

The Banded Rail is one of those enigmatic species that just penetrates the Oriental region in the no man's land bordering Wallace's Line. It is quite widespread through the islands that extend eastward as far as Australia. No less than 24 distinct subspecies have been described, many inhabiting only a single island.

The distribution of several species of Oriental birds (below) shows a curiously disjoined effect, with individuals in the eastern Himalayas and in the Western Ghats and Sri Lanka. The Great Pied Hornbill and the Asiatic Broad-billed Roller are cases in point.

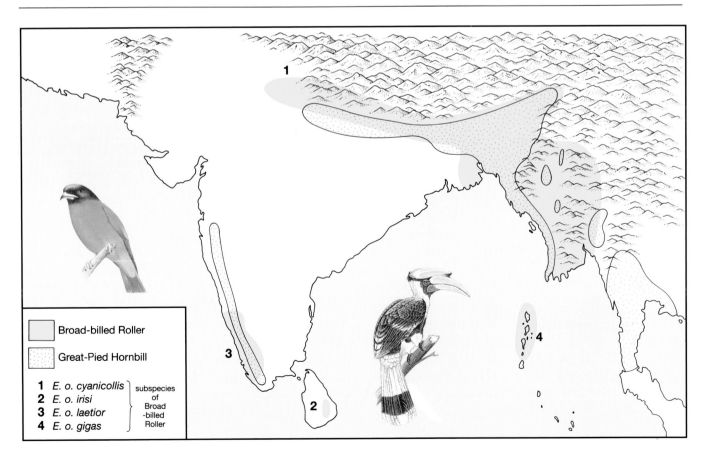

Broad-billed Roller

Great-Pied Hornbill

1 *E. o. cyanicollis*
2 *E. o. irisi*
3 *E. o. laetior*
4 *E. o. gigas*

subspecies of Broad -billed Roller

The systematics of the Old World paradise flycatchers has strained the minds of ornithologists for generations. Today the genus Terpsiphone *is divided among 12 species. Some, like the Seychelles Paradise Flycatcher, are confined to a single isolated island. Others, like the Asiatic Paradise Flycatcher, shown here feeding its young, range from Arabia to Sumatra and have been divided among no less than 18 distinct subspecies. A male may take six years to acquire full adult plumage and is then a ghost-like white apparition that flits among dark forest glades.*

again in southern India and Ceylon. Called the "Satpura hypothesis" by S. L. Hora, a distinguished Indian zoologist, this disjointed distribution of birds, separated by over 1,000 miles of nonoccurrence, is explained by the presence of a once continuous mountain system from the eastern Himalayas across peninsular India to the Satpura mountains, Kerala and Ceylon.

Various species and species groups are now found only at the two extremities of this "chain". They include some of the laughing thrushes, Fairy Bluebird, Great Hornbill, Broad-billed Roller and Malabar Trogon. In some cases these isolated populations have evolved subspecific status, but in others even such slight differences are not apparent. Strangely enough, it was the distribution of freshwater fish that led Hora to his hypothesis.

One interesting, and perhaps relevant, aside is the remarkable absence of vultures from Ceylon. These large, powerful fliers are abundant over much of India, and in the past Ceylon has certainly been joined to the "mainland". The leopard colonized from the north, but earlier than the tiger, which arrived in India after Ceylon had become an island. Consequently, the leopard is the dominant predator in Ceylon and hunts by day. In India, however, it is wary of the tiger and is essentially nocturnal as a result. Yet the vultures could, even today, make the short crossing of the Palk Strait with little difficulty and there is certainly sufficient food in Ceylon to maintain a healthy population of scavengers.

Although most of the Oriental region lies in the northern hemisphere it still acts as winter quarters for birds from the eastern half of the Palearctic. Ducks and geese pour into favored areas, sometimes in enormous numbers. Waders, too, may be abundant, leaving their summer homes among the tundra to find refuge along the coasts of India, Malaysia and Indonesia. Some, such as Wood Sandpipers, seek wintering grounds among inland marshes and arrive, in India for example, at the end of the monsoon when the wetlands are full and extensive. As the landscape dries out they find alternative sites around the ponds that are an integral part of village life.

In the western part of the region many wader species are familiar to the European visitor. Further east the birds are less familiar and more sought after as a result. The Little Stint, for example, extends no further east than India and Bangladesh in winter, whereas Red-necked and Long-toed Stints are found no further west than the east coast of India. The singular Spoon-billed Sandpiper of Pacific Siberia winters only in the Oriental region, as does the Spotted Greenshank of northern Japan. Long-distance migrations of these shorebirds are paralleled by similar migrations among the smaller passerines. In this manner, the whole population of Lesser Whitethroat moves from Europe southeastwards to winter among the plains of India and adjacent areas. Greenish and Arctic Warblers head in the same direction, as do Red-flanked Bluetails and Siberian Rubythroats.

The Himalayas offer a secure sanctuary to the Lammergeier that flies high above the mountain fastnesses of the Everest area. Only a very few mountain areas of the Old World now hold this magnificent bird that feeds largely on bones left behind by other scavengers.

149

Right and Opposite: *The White-bellied Sea Eagle is one of the three sea eagles to occur in the Oriental region. While two derive from the north, this particular species is clearly of more easterly origin, being numerous in Australia. It is a large and powerful predator with grasping toes and long, sharp talons, ideally suited to capturing fish.*

It is generally agreed among ornithologists that transcontinental migrations such as these indicate the direction from which birds have colonized fresh breeding grounds; which, in turn, gives us a major clue to their area of origin. Thus, for example, a genus of small leaf warblers, the *Phylloscopus*, such as the Greenish and Arctic Warblers mentioned above, heads southeastwards in autumn toward India and China. While in itself this may not be sufficient to say that these species originated in this part of the Oriental region, when coupled with the distribution and migration pattern of other leaf warblers we can say with confidence that the genus does indeed have its origins in the Orient. The fact that so many members of this fascinating genus can be found in India, the Himalayas and China makes the region particularly attractive to the European bird-watcher.

The Oriental region is also a winter home to large numbers of Palearctic birds of prey and to eagles in particular. Nowhere is this more obvious than among the marshes of the Bharatpur reserve in northern India where these great birds are often surprisingly abundant. Steppe Eagles are the most common, but Spotted and Imperial Eagles as well as the White-tailed Eagle are all regularly present. Add in the resident Tawny Eagle and Pallas's Fish Eagle and the number of species is really impressive.

Usually these great birds of prey simply sit on top of a low tree and watch the passers-by, for they are unmolested and exceptionally tame. They hunt

at first light and by the time even the early rising bird-watcher is about they have full crops. With the exception of the two "fish" eagles, these are difficult birds to identify, with a complex succession of plumages as they become mature. There is thus considerable confusion among visiting bird-watchers and claims for birds that are either not present or decidedly thin on the ground. On my first visit to Bharatpur I was informed that Spotted Eagles could be seen sitting around virtually everywhere. In fact the vast majority turned out to be Steppe Eagles.

The presence of so many well-satisfied birds of prey at the sanctuary at Bharatpur needs little explanation. Once the hunting grounds of the local maharaja, the shallow pools act as a magnet to a huge array of birds, many of which are present in large numbers. Geese, ducks and waders migrate here in their thousands to join the vast numbers of storks, ibis, herons and

Many of the birds of the Oriental region are fortunate to live alongside a human population that values life, in all its different forms. The birds are generally unmolested in India, though persecuted in Thailand and elsewhere. The Painted Stocks of Bharatpur are symbolic of the special relationship between man and birds in India.

egrets that breed. There are literally birds everywhere and 150 species seen in a day is a relatively simple exercise. Of course, there are concentrations of water birds elsewhere in the world, but even the best traveled birders regard Bharatpur as something very special, a unique sanctuary in the midst of the plains of India.

Bharatpur shows what can be achieved in an area that is teeming with people in one of the most densely populated countries of the world. The lesson is that birds are able to take advantage of even a tiny oasis if the habitat is suitable to their needs and they enjoy a modicum of protection and peace. It is a lesson that has applications worldwide.

If this brief summary of the birds of the Oriental region has shown us anything, it is the complex nature of a region that abuts so many others. We have seen the difficulties of distinguishing one fauna from another in the area between Indonesia and New Guinea. We have seen how the mighty Himalayas may form a barrier both to colonization and migration on a north–south axis to some birds, but not to others. And we have noted the effect of man on birds in this, the most densely populated region on earth.

CHAPTER SEVEN

AUSTRALASIA

THE AUSTRALIAN REGION

The Australian region, naturally, includes the great island continent of Australia together with its two large island neighbors – New Zealand and Papua New Guinea – as well as the myriad of islands of the Pacific. With no land boundaries with neighboring regions, it would be easy to think that this might be one of the most clearly defined of all the world's zoogeographic regions. But, as we have already seen, there are considerable problems in drawing a boundary in the northwest among the islands of Indonesia where the zone of transition is now referred to as "Wallacea". Those who have read this book in out-of-order chapter sequence (and why not?) should turn to the early part of the chapter on the Oriental region for a fuller account of this fascinating part of the world.

Boundaries of the region are considerably easier to delimit in other directions, though none are without interest or question. With the breakup of the great Gondwanaland supercontinent 200 million years ago, Australia remained firmly joined to Antarctica. Indeed, it became separated less than

The Common Melipotes, or Smoky Honeyeater, is found only among the forests of New Guinea where, as with so many other species, much of its life is shrouded in mystery. It is the "newness" of much of the Australian region, that makes it such a fascinating part of the world for birds.

Opposite: *A kaleidoscope of color as Rainbow Lorikeets come to a feeding station near the coast of northeastern Australia where they are encouraged as a tourist attraction.*

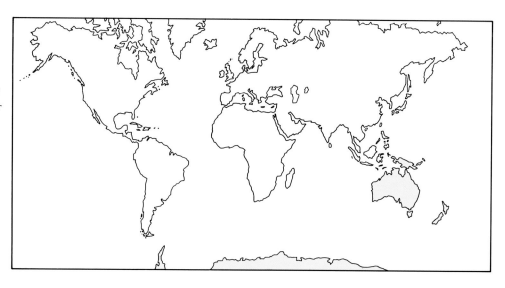

A pair of Tawny Frogmouths heavily camouflaged among the branches of a eucalyptus in Queensland, Australia. This small family of nightjar-like birds has its epicentre in the Australian region, though some species have spread westwards into the Oriental region. Their nearest relatives would appear to be the potoos of South America.

65 million years ago at a time when, as all authorities agree, birds were well developed as a class of animals. Many therefore regard the Australian and Antarctic regions as one and, to avoid debate in what is at best a general account of the world's birds, we shall treat them here as no more than subregions. However, while Australia broke away from Antarctica in one direction, New Zealand did so in another. Thus, while now separated only by the comparatively narrow Tasman Sea, these two landmasses have their origins thousands of miles apart.

In the east the Australian region extends right across the Pacific almost to the coasts of Mexico and Chile where the boundary is drawn by a straight line just to the west of the Galapagos Islands. Northwards it extends as far as latitude 30° north and includes the islands of Midway and Hawaii. Southwards the boundary approximates to latitude 50° south where it is drawn, somewhat variably, among the sub-Antarctic islands to the south of New Zealand where "Wallace-type" faunal distinctions can be made.

The region thus straddles the equator, though the great landmasses all lie well to the south, between the equator and 50° south. By far the largest land area is Australia itself covering nearly three million square miles. Papua New Guinea in comparison covers under 200,000 square miles and New Zealand a mere 100,000 square miles. Yet despite this remarkable difference in size both Australia and Papua New Guinea can claim some 570 breeding

species each. Even the comparatively tiny New Zealand has a list of nearly 220 breeding species. Altogether the Australian region can boast a total of 900–950 breeding species, making it, with the Palearctic, only the fourth richest of the world's avifaunas.

If, however, we regard "richness" not in terms of absolute numbers but in terms of "uniqueness", then the Australian region achieves a completely new status. For of the 570-odd Australian breeding species, no less than 360 are endemic. It is doubtful if such a high level of endemism can be found anywhere else on earth within the boundaries of a single country. At a different level, the Australian region also boasts a considerable list of endemic families and, of course, even more genera. These include, among others, the Emu, cassowaries, the remarkable megapodes, frogmouths, lyre-birds and bowerbirds, all of which are totally absent elsewhere in the world. Indeed, in most cases, it is impossible to establish which other groups of birds are even vaguely related to these Australasian specialities.

Nevertheless, many elements in the avifauna do have clear relatives in the Oriental and Palearctic regions of the Old World, the nearest of the world's other great faunal regions. Grebes, rails, cranes, hawks, kingfishers and crows are all widely present, while the Cattle Egret is the most recent colonist. Despite such connections, there are also some quite notable absentees. Flamingoes, vultures, sandgrouse, trogons, barbets and wood-

The New Zealand Grey Flyeater is one of those endemic New Zealand birds that has managed to survive despite the assaults of introduced alien species. Its near relatives are spread through the Australian region and it doubtless originated from colonists from the larger landmass.

The Red-eared Firetail is a well-marked finch of southwestern Australia that has been hard hit by housing and other developments that have affected the forested glades in the area around Perth that it prefers. Unlike most other small finches it is seldom, if ever, found in flocks and is easily overlooked as a result.

peckers, finches and buntings have all failed to establish themselves in the region. In many cases Old World species have only just managed to cross into the transitional zone ("Wallacea"). In other cases the penetration has been strange. Blyth's Hornbill, for example, is widely distributed in that "hornbill paradise" that extends from Burma through to Indonesia, but is then the only member of the family from New Guinea eastwards to the Solomon Islands. Similarly, the widespread Old World genus of *Turdus* (thrushes) is established in New Guinea, some of the Pacific islands and at Lord Howe Island in the Tasman Sea, but is absent from Australia.

Such an overall "insularity" in terms of isolation from neighboring regions is also obvious in the form of migrants. A wide variety of northern waders spend the winter in the region, but there is an almost complete lack of migrant small birds. Virtually all the warblers, swallows, chats, wagtails, swifts, flycatchers and so on are completely absent. This lack of penetration of such an obvious rich wintering ground is particularly remarkable when compared with the wealth of species and numbers that spend the lean season in those other large southern hemisphere landmasses, Africa and South America.

Within this vast region three distinct subregions are reasonably well defined and worthy of more than a casual mention. While Australia is quite definitely the "base" from which much of the fauna of Papua New Guinea

The Brown Kiwi (below), New Zealand's most famous endemic bird, hunts by dark among the litter of the forest floor. Decimated by introduced pests, it survives in small numbers as at Rotoroa, where this bird was photographed. Having dispensed with flight the Kiwi is able to carry a far larger egg than any other bird, up to 30 percent of its body weight, as seen in an X-ray (left).

A male Victoria Riflebird calling from a forest stump in northern Queensland, Australia. This is one of only a small group of birds of paradise found outside New Guinea. It is a solitary, forest-based bird marked by irridescent patches on an otherwise black body. The bright yellow mouth may be an effective signal in its dark forested habitat.

derived, the latter subregion shows a strong development of endemic species. Thus the fantastic birds of paradise are almost entirely confined to Papua New Guinea, with only a few species present in adjacent Australia.

In the same way, New Zealand owes most of its birds to Australia, its endemics being no more than forms isolated from their parental Australian stock. For instance, the various New Zealand wattlebirds are derived from successive invasions of these islands by Australian Apostlebirds. The owl-parrots result from similar invasions by the Night Parrot of Australia, which is fast becoming extinct. In isolation such invasions would doubtless lead to a radiation of species similar to that of the Hawaiian honeycreepers or Galapagos finches, in which a single invader develops into a variety of distinct forms to occupy a wide range of vacant ecological niches. That this has not happened in New Zealand seems to be because invasions of the original form continue to take place. The only New Zealand family to show adaptive radiation are, in fact, the New Zealand wattlebirds – the Saddleback, Kokako and the now extinct Huia.

One further group of New Zealand endemics are the ratite groups of moas and kiwis. These flightless birds can hardly have enjoyed successive invasions of the islands and are usually considered as having derived from the Australian Emu. We have, however, examined the case of the world's largest flightless birds at some length in earlier chapters. Briefly, there is a case for considering that all of these species may have descended, not from flying ancestors, but from a nonflying ancestral stock that once roamed Gondwanaland before the great southern continent split apart.

The fact that the moas are now extinct should not detract from the fact that the Australian region is particularly well endowed with ratite species. With three species of cassowary, an Emu, three species of kiwi and anything between six and 26 species of moa, the region has several times more species of these birds than the rest of the world combined. It is easy to explain that with the absence of large terrestrial predators large flightless birds are more likely to have survived in Australasia than elsewhere in the world. But this does not explain the predominance of these birds here as against other similar areas – particularly when it comes to fossil forms.

It is in New Zealand that the ratites reached their highest level of development. The only extant species, the kiwis, are unusual for their extraordinary life-style – virtually blind, covered with hair-like feathers and with a long decurved bill equipped with highly sensitive nostrils located at the tip. They are also unique in laying the largest egg in proportion to body weight of any of the world's birds – approximately one third.

The moas, however many species there once were, are now all extinct. Some, at over ten feet in height, were the world's tallest birds and second in weight only to the also extinct elephant birds of Madagascar. From the evidence that has been gathered there can be no doubt that large flocks of moas, of perhaps 19 distinct species, once roamed the grasslands and open woodland of New Zealand. Their demise was doubtless due to their culinary qualities, as successive waves of people from Polynesia settled the land. Certainly most, if not all, of the largest species were already extinct by the time Europeans arrived.

Reasonable descriptions of large flightless birds on South Island were obtained from sealers who spent months living off the land in the remoter southern parts of the island in the 1850s. And in about 1880 Alice McKenzie, then seven years' old, saw a moa-like bird just over three feet tall. Her accounts are convincing enough for us to conclude that she was probably the last person ever to see a live moa.

Another group of endemic, ground-dwelling birds are the megapodes, or mound-builders, three species of which are found in Australia. These are medium to large birds, related to the pheasants and other gamebirds that are not otherwise particularly well represented in the region. While the Scrubfowl is confined to the northern coast and the Brush Turkey to New South Wales, the Malleefowl is more widespread through the southern half of Australia. The latter is unique in occupying open scrub (called mallee) rather than the dense forests occupied by other species.

All megapodes share the strange habit of burying their eggs and allowing natural heat to perform the incubation for them. Tactical differences

The Australian region boasts several of the world's most construction-orientated birds. These Malleefowls (right) are tending the huge pile of plant debris and earth that acts as an artificial incubator for their eggs. Building and tending the mound is virtually a year round occupation for these birds. The male Satin Bowerbird (opposite) has built its bower, an avenue of twigs decorated with specially chosen objects, to impress visiting females.

between the species vary enormously, however. In rocky areas each egg may be laid in a crack where the heat of the sun and the cold of the night are balanced by the heat-retaining rock. Elsewhere a hole may be excavated in sand subject to warming by volcanic steam. But most megapodes create a mound of soil and vegetation where a sort of compost-heap effect warms the eggs constantly.

Living in semidesert country, the Malleefowl faces such extremes of temperature between day and night that it has adapted a quite unique method of mound control. Both male and female Malleefowl combine to excavate a pit some 16 feet across and three feet or more deep. With a general lack of vegetation in mallee country, leaves are gathered systematically over a large surrounding area and brought to the pit. Following rain, this layer of leaves is covered with a thick layer of soil. The eggs are laid in a pit in the center and are warmed by the decaying vegetation below. Because the heat generated varies so much in such an open environment, the male bird checks the temperature throughout the incubation period. Sometimes he will remove soil to allow the heat to escape, and at other times he will add soil to increase insulation. During cold periods he will almost expose the eggs to the heat of the sun in the morning and rebury them again late in the afternoon. Altogether the male Malleefowl spends practically his entire year either building or tending the mound. When the eggs hatch the young, which are well feathered, dig their own way out and run swiftly for cover among the surrounding bushes. They are capable of flying within a day.

The bowerbirds, another endemic family, are also great constructors, though in a quite different way. The males of these forest-dwelling birds are mostly brightly colored and spend much of their time creating special courting grounds on the forest floor. These are carefully decorated with a variety of objects, again brightly colored, that are placed with the greatest of care for maximum effect. While some species show a distinct preference for a particular color, others seem to prefer shiny objects. Some even use crushed berries to paint their bower, a remarkable piece of "tool-using", unique in the world of birds. It is difficult to understand why birds that are highly colored and highly vocal, with superb mastery of mimicry, need to go to such lengths to attract a mate. However, female bowerbirds, which are generally rather dully colored, do need a lot of persuading and, although they regularly attend their mate's bower, they are only prepared to mate if the available food supply is sufficient to rear their brood.

Two quite distinct types of bower are constructed. The best known, that of the dark, shiny-blue Satin Bowerbird, is an avenue of sticks planted on the ground and decorated with objects as varied as feathers and bottle tops.

This Superb Bird of Paradise spreads its "cloak" of bright blue feathers as it begins a display. These birds, like most other members of the family, are confined to the forests of New Guinea where the attention of females is attracted by loud calls.

Usually it prefers blue objects and these may be gathered in variety and quantity. It will often paint the twigs of its bower with berry juice, occasionally using a piece of bark or other object as a paintbrush. Some authors consider this painting as a misplaced form of courtship feeding, in which the food is offered to the bower rather than to the female. Such is the male's attention to detail that objects that have been misplaced will be returned to their correct position, accompanied by a great deal of angry clucking. The Tooth-billed Catbird does not construct a bower at all, but clears an area of forest and then places upturned leaves to cover the ground. As an experiment, on one occasion the leaves were carefully turned over to show the shiny upper surface. When the owner returned he immediately turned them all the "correct" side up again.

The collection of materials by avenue builders may extend over a considerable area. One bower of a Spotted Bowerbird contained hundreds of nails that had each been carried over 400 yards from a barn being repaired on a nearby farm. Probably the most spectacular of bowers are the maypole-type in which a collection of twigs is placed around the base of a sapling. In the case of the Golden Bowerbird this may create a maypole nearly ten feet in height, one of the most extraordinary feats of construction

A male Red-winged Parrot feeds on grevillea seeds. This is a common and gregarious species found in northern and eastern Australia and in southern New Guinea, that is predominantly arboreal in habit. It is frequently found near water and, like so many other parrots, nests in tree-holes.

in the animal world. The Brown Bowerbird, in contrast, builds his bower to no great height but then constructs an umbrella of twigs at the top.

Having constructed a bower and decorated it to perfection, the male calls loudly to attract his mate. Should she put in an appearance he will display vigorously, often picking up some of his collection of apparently irresistible objects to show them off. He will not, however, approach the female, but concentrates his attention on the bower. Such gentlemanly behavior is perhaps typical of the dandy.

Closely related, but placed in a quite separate family, are the equally dramatic birds of paradise. With only a few species in Australia, these birds are otherwise confined to the island of Papua New Guinea where they have proliferated into a wide variety of extraordinarily beautiful forms. The name stems from their first arrival in Europe when the explorer Magellan showed them to an astonished Spanish court who, quite reasonably, thought they must have originated in paradise. An alternative, and no less plausible, explanation is that, when they were first imported, the skins were lacking legs and feet and so the birds were presumed never to perch or come to ground and thus to derive from paradise. Whatever the explanation, the trade in their fantastic feathers boomed throughout the nineteenth century

Above: *A flock for Budgerigars comes to water to drink. The wild Budgerigar is always a green and yellow bird, unlike the varied domesticated forms that have been specifically bred to produce different color forms.*

Right: *The Kea is an exceptional parrot that has evolved the habits of a scavenger among the mountains of New Zealand. Its curved bill, designed for eating fruit, is more hawk-like and it regularly feasts on dead animals.*

with disastrous results for the wild populations. Fortunately, ladies' fashions changed before too many birds were destroyed and the world's most beautiful and colorful birds survive in their native habitats.

Birds of paradise are forest-dwellers, many of which inhabit the remote interior of Papua New Guinea. Despite settlement and exploitation, vast areas of this island remain both remote and inhospitable. Here primitive tribespeople hunt them for the plumes that are still widely used on ceremonial occasions. Male birds not only have beautiful plumes and are boldly colored, but also perform elaborate displays among the forest trees. Some actually hang upside down from branches, the better to show their finery. Though they are only small to medium in size, many have highly extended feathers that grow from flank, nape or tail. The King of Saxony Bird of Paradise, for example, has two broad plumes two feet in length that are much prized by local tribesmen who wear them through their noses. The Ribbon-tailed Bird of Paradise has tail feathers over three feet in length. When first discovered in 1938 it was described only on the basis of these feathers found in the headdress of a tribesman.

Despite being such a familiar family, these birds remain little known. Few have been photographed or studied. The details of their lives remain largely secret and many nests and eggs have never been described.

The Australian region is particularly well off for another group of colorful birds – the parrots. These are mostly gregarious open-country birds, particularly likely to descend on some waterhole or favored feeding

Birds of dense forest have to ensure that they can be found by prospective mates and then that they are suitably attractive. The superb Lyrebird is one of the world's greatest mimics and its loud calls echo through the rainforests of eastern Austalia. When a female tracks him down, the male spreads its Peacock-like tail to attract her.

ground in huge numbers. Notable in this respect is the Budgerigar, a green-yellow bird in Australia but which has been widely bred in captivity to create a range of color forms. Their propensity to gather in huge numbers has been commercialized in parts of eastern Australia where large flocks of Rainbow Lorikeets descend daily on bird gardens providing food. The birds are so tame that they even use tourists as perches.

Parrots have two toes pointing forward and two back, an arrangement that, like the woodpeckers, is ideally suited to climbing trees, where most species nest in holes. The bill is sharply hooked, like that of a bird of prey though, in this case, fruit rather than flesh is the predominant food. However, so similar are the bills of these two totally distinct groups of birds, that one parrot, the Kea of New Zealand, has actually transferred its attention to carrion. Another New Zealand parrot, the Kakapo, has become virtually blind and flightless like a Kiwi and is now seriously endangered, like so many other island endemics faced by introduced predators.

On the other hand, most parrots are fast and strong fliers. The wings are pointed and the tail is frequently marked by extended central feathers. While Australasia is well endowed with these birds, they are equally numerous in South America. Despite being placed in six distinct sub-families, groupings of the world's parrots above generic level is decidedly tenuous. Parrots are all closely related, no matter where they are found. Just how they originated remains a mystery, unless, that is, they originated before the great continents drifted apart.

A male Common Cassowary with his two well-camouflaged chicks. These large flightless birds are fearsome fighters when disturbed or attacked and have managed to survive the onslaught of introduced predators such as the dingo, or wild dog. The chicks, in contrast, are highly vulnerable and rely on their parents and coloration for protection.

While we have concentrated much of this chapter on the unique and the typical elements of the larger landmasses of the Australian region, it would be quite inappropriate not to say something about the myriad of islands that extend eastwards across the Pacific. There is certainly good evidence to show that most of these are inhabited by birds that colonized from the west, from Australia and Papua New Guinea. The further east one goes the fewer species one finds on each island. Thus the Solomon Islands certainly have an impoverished avifauna when compared with Papua New Guinea, even though they are no great distance east and are virtually linked by other relatively large islands. Taking all of the southwest Pacific from the Solomons to Fiji we find a global total of 388 species occurring compared with 650 for New Guinea alone. These totals include many migratory seabirds and waders which, if eliminated, would make the totals even smaller. The Solomon Islands boast a list of 148 species of land and freshwater birds, and New Caledonia with 68 species is the richest east of the Solomons. By the time one reaches Samoa the total has dropped to little more than 30 species, and still further east numbers continue to drop.

The decline in the number of species eastwards through the Pacific therefore shows that the origin of the birds of these far-flung islands was essentially Papua–Australian. Inevitably, the more isolated island groups are more likely to have developed their own endemic species and the process of species radiation reaches its peak among the islands of Hawaii, more or less halfway between California and Papua New Guinea. Here, in particular, we find the paradigm radiation example – the sadly depleted Hawaiian honeycreepers. We have already considered this remarkable group of birds in the first chapter.

Geographical isolation is quite definitely the causal factor of speciation, but such isolation need not be measured in either thousands, or even hundreds of miles of open sea. Much depends on the nature of the colonizer itself. A strong flier, for example, can easily hop from island to island and never become isolated. Conversely, a weak flier may remain isolated within a short distance of a neighbor. For this reason the Pacific is littered with endemic crakes and rails found only on single islands. Sadly, there is also a tendency among rails to become flightless when faced with a lack of predators. I say "sadly" because the nineteenth and twentieth centuries have seen the spread of commercialism throughout the whole world, including the Pacific. This, in turn, has resulted in the deliberate and accidental introduction of alien predators such as cats, rats and mongooses and the consequent demise of many unique birds. No less than 14 species of rail have become extinct in the Pacific, ranging from the tiny Laysan Rail to the huge Barred-wing Rail.

Though not related to the rails, the Kagu of New Caledonia seems likely to follow the same route to extinction. This substantial bird, some two feet long, was discovered only in the 1850s and was at first thought to be a heron or egret. It was only later that it was classified as a member of the Gruiformes, an order that includes the Sunbittern of South America. Nevertheless, the Kagu is only superficially similar and is now regarded as a separate monotypic family. The birds live among the mountains of New Caledonia, a large island to the east of Australia roughly halfway to Fiji. Its feathers are loose and hair-like and an overall gray in color, and its bill and strong feet are red. The Kagu is completely flightless, a factor that remained unimportant until New Caledonia was settled about the middle of the nineteenth century. Cats and rats doubtless had an immediate effect, especially on eggs and young, but the importation of dogs about 1860 set the Kagu into an immediate decline. By 1940 it was so rare that numbers were being trapped for export. Today it is found only in the remote mountains of the southern half of the island and is extremely rare. Fortunately, these birds do well in captivity, being both long-lived and active breeders. So as long as a good stock ensures a broad gene base there will always be the chance of a reintroduction program within a well-fenced "wild" enclosure. It is quite clear that reintroduction to a land filled with predators would be an unqualified disaster.

If the Pacific, as such, has a history of extinct and all-but-extinct birds, so, too, does New Zealand which to all intents and purposes we can regard as two large Pacific islands. As we have seen, the New Zealand avifauna basically originates in Australia and has been topped up over the centuries by fresh waves of immigrants that prevented species radiation. Indeed, the only family to have developed within New Zealand is the wattlebirds. Of the three species, one has already become extinct and that was unfortunately one of the world's most extraordinary birds. The male and female Huia

Though penguins, like these Chinstraps (left), are often seen on icebergs, they do not nest on the ice. Instead they gather at ice-free areas, like these King Penguins (below) at one of their South Georgia rookeries. In a short time the area turns into a muddy quagmire, made all the more unpleasant by the stench of rotting fish and bird droppings.

A Snow Petrel at its nest among a scree of the mountains that protrude from the Antarctic ice cap. These all-white petrels are probably the world's most southerly breeding birds, but they inhabit a virtually predator-free zone as a result.

were virtually identical. Both were 20 inches in length, black with an irridescent sheen of green, and both had bold red wattles on their cheeks and white tips to the rounded tail feathers. They differed only in the structure of their bills and their feeding methods. As this is often a major factor in the evolution of species – that is, different feeding methods inhibiting competition – it is extraordinary that the male and female of a single species should show such sexual dimorphism. The male Huia had a robust decurved bill; his mate had a longer and thinner probing one. And while the male hacked at decaying trees to obtain grubs, the female would probe into their tunnels to extract grubs. The pair often hunted together, however, when the female would probe among the tunnels exposed by the male's more robust technique – although there is no evidence of food obtained in this way being consumed other than by the female.

Being arboreal in habit Huias were not defenseless against introduced predators as were the rails and other ground-dwelling birds. In their case it was the size and significance of their tail feathers that led to direct human persecution. For although Maoris had worn them for generations it was only the increased demand from Europeans that seriously reduced the Huia population. Once decline had set in the specimen-collecting ornithologists quickly polished off the remainder. The last report was in 1907 and, most regrettably, it seems unlikely that such a large and obvious bird could have survived unobserved for the best part of a century.

To round off our story of destruction among the islands of the Australian region we can return to Hawaii where the 22 species of honeycreeper not only provided the best examples of adaptive radiation, but also the best example of human destruction. Only 12 species are known to be still present on the islands and most are extremely rare. Taking the avifauna of Hawaii as a whole, of the original 68 species or subspecies of endemic birds no less than 40 are now either extinct or on the danger list.

This then is the end of a story of a wealth of species in a zoogeographic region that was largely untouched by man until a mere 200 years ago, and of what disasters can happen when a fragile world is treated without care. It is a depressing story but, hopefully, one from which we can learn something – though I very much doubt it. Should any reader wish to see anything of the remarkable bird life of the Australian region I would suggest an early journey before only the most adaptable and opportunist species are left. The endemic birds of New Zealand are already singularly difficult to see; Papua New Guinea will probably follow suit, along with the scattered islands that so many once regarded as an earthly paradise.

The Antarctic Subregion

Although the Antarctic is referred to as the last great wilderness left on the planet and, as I write, the subject of international agreement on its exploitation for at least another 50 years, it cannot properly be claimed as a full faunal region. In fact, this vast continent is covered by such a weight of ice that much of it is below sea level. Only here and there along the coast can land be seen, and even then, only during the brief weeks of summer. The continent, which reaches furthest north in the Graham Land peninsula, is surrounded by sea and is easily delimited. It does, however, have many satellite islands that should be treated as lying within the region.

Because all birds, bar one, come to land to breed, the Antarctic is a barren region, poorly off for species. Taking in every offshore island, only some 90 different species have ever been recorded and less than 40 have bred. Indeed, all Antarctic birds are either dependent on the surrounding sea or on other birds or mammals that are themselves dependent on it. By far the most abundant and obvious are the seabirds themselves, birds that only require land as somewhere to lay and incubate their eggs and rear their young. For them the surrounding oceans are rich in fish and other marine life, so Antarctica is no more than a gigantic nesting site.

Only two birds have managed to use more than the coastal strip of the continent. The Snow Petrel nests among the rocky mountain peaks that jut up from the sea of ice up to 150 miles inland. Emperor Penguins are equally unique in actually nesting on the ice itself.

Many other penguins breed around Antarctica, but relatively few on the continent itself. In addition to the Emperor we can add the Adelie, the Gentoo and the Bearded Penguins. Only another ten species have been recorded breeding in Antarctica itself, including the Cape Pigeon (actually a petrel), Dominican Gull, South Polar Skua and the Antarctic Tern. All the other 25 or so birds that breed do so on the offshore islands. Among these none are so abundant and diverse as the petrels and albatrosses.

Above: *A Light-mantled Sooty Albatross alights near its South Georgia nest site.*

Although we think of penguins as being Antarctic birds, only 6 of the 18 southern penguins actually breed on the sub-continent itself and only two of these, Emperor and Adelie, are confined to the sub-continent. Other species are more widespread, though the Jackass is only found along the coasts of South Africa and the Snares' Island Penguin is only found on the island of that name off southern New Zealand.

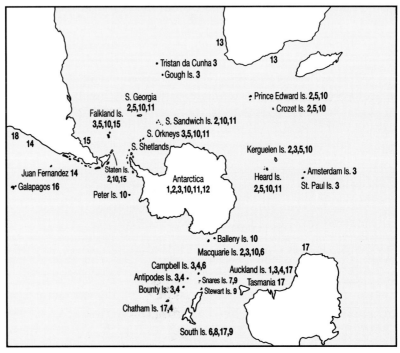

1	Emperor
2	King
3	Rockhopper
4	Erect-crested
5	Macaroni
6	Royal
7	Snares Island
8	Fiordland Crested
9	Yellow-eyed
10	Gentoo
11	Chinstrap
12	Adelie
13	Jackass
14	Peruvian
15	Magellanic
16	Galapagos
17	Little Blue
18	Humbolt

Of the world's 100 or more species of these "tube-nosed" birds, no less than 65 breed around Antarctica. They vary from the huge Wandering and Royal Albatrosses to the diminutive storm petrels. There is considerable evidence that some of these great ocean wanderers actually circle the earth outside their breeding season.

At their breeding colonies many seabirds gather in extraordinary numbers. Penguin colonies, for example, quickly turn into a muddy quagmire, with a smell that is both unusual and foul. It is at these colonies that scavengers like the sheathbills and Dominican Gulls gather when they are not busy clearing up the mess created at seal colonies. Here, too, the piratical skuas are much in evidence, taking eggs and chicks as well as hustling adults with food. Ornithologists are divided about the systematics of the "Great Skua" group, with most regarding the northern Great Skua as quite distinct from the birds of the Antarctic. Others go further and divide these southern skuas into several distinct species. Most now agree that there are two, perhaps more, species overall – the Great Skua in the north and the South Polar Skua in the south.

Antarctica is, then, unique in many ways. It is home to several endemic birds, but to no endemic families; and it has more of the world's albatrosses and petrels than any other region, but only supports a handful of land birds. Nonetheless, despite the words and film lavished upon them, the birds that live here are still relatively unknown.

Wandering Albatrosses flap the longest wings in the world in display. The Antarctic, together with the surrounding islands and seas, is the world headquarters of the great albatrosses and larger mollymauks. These long-ranging seabirds use the updraft of air created by large rolling waves to keep them airborne and provide the power for gliding movement. Curiously, although albatrosses have managed to colonize the North Pacific, they are all but unknown in the North Atlantic.

CHAPTER EIGHT

SOUTH AMERICA

THE NEOTROPICAL REGION

South America is home to more species of birds than any other part of the world and can be regarded, therefore, as the Mecca of ornithology. Two South American countries, Colombia and Ecuador, each boast a list of over 1,500 species, a total that compares favorably with every other zoo-geographic region taken as a whole. Just why this continent should be so richly endowed we shall examine later. For the moment, suffice it to say that a combination of physiology and climate have combined with geographical isolation to produce the most extraordinary species radiation within what are, in many cases, endemic or near endemic families.

The Neotropical region is relatively easy to delimit. In the east it is separated from Africa by the South Atlantic, an ocean virtually devoid of islands that might provide a link between these two great southern landmasses. To the west the huge Pacific ocean separates it from Australia and from both Palearctic and Oriental regions. Though the Pacific is well stocked with islands, most are located well to the west and only a few are

*The Hoatzin forms not only one of the Nearctic region's endemic families, but also shares a lifestyle similar to that postulated for the world's first known bird –
Archaeopteryx.*

Opposite: *The toucans, such as this Chestnut-eared Aracari, are a typical Neotropical family of fruit-eating, forest birds. They are replaced by the hornbills of Africa and Asia, which have similarly over-sized bills.*

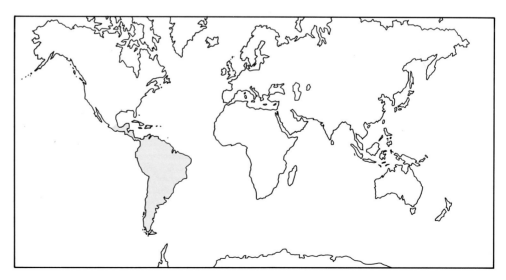

This Emerald Toucanet has chosen a tight-fitting nest hole in which to incubate its eggs and rear its young. Though photographed in the cloud forest of Chiapas, Mexico, this species ranges southward through Central and South America as far as Peru.

situated close to South America. It is interesting to note that while man is regarded as having colonized the Pacific from the east, birds have generally arrived from the west. A notable exception is the remarkable avifauna of the Galapagos, but these islands lie less than 600 miles off the coast of Ecuador.

In the south the Neotropical region is separated from Antarctica by the relatively narrow, but still formidable, barrier of the Drake Passage. But at 58° south, the island of Tierra del Fuego is already an inhospitable place for birds and only a relatively few species are found on either side of the dividing sea.

Only in the north does the Neotropical region share a land boundary with another zoogeographical region, but even here the land link is of relatively recent origin. The last water barrier did not close until the Pliocene period, some 3–12 million years ago, so that much of the Neotropical avifauna was able to evolve in isolation. Today the boundary is drawn through Mexico at the northern limit of the tropical rainforest, about the latitude of Mexico City. Inevitably there is an area of overlap between the Neotropical and Nearctic regions in that country and elsewhere in Central America, but while the present region boasts an avifauna in excess of 2,500 species, its northern neighbor can summon up a mere 750 species.

What then are the factors that make the Neotropical region such a paradise for birds? Firstly, its position covers an area extending from 24°

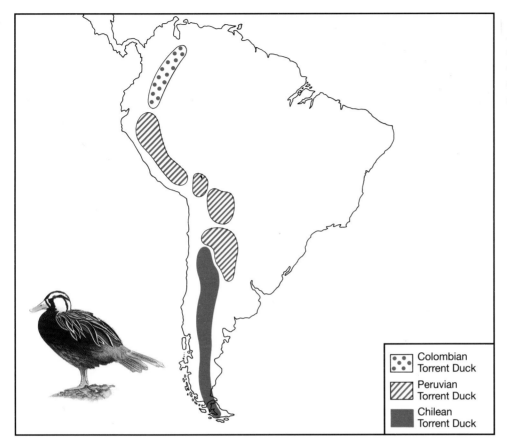

Colombian
Torrent Duck

Peruvian
Torrent Duck

Chilean
Torrent Duck

Torrent Ducks are confined to the Andes and, despite plumage differences, are generally regarded as forming a single species. They differ mainly in the color of the male underparts – ranging from virtually white to pure black.

north to 55° south, a total of some 80° of latitude and greater than any other region. Secondly, it boasts the world's greatest mountain chain, the Andes, which extends the full length of the continent. Thirdly, it has less desert – a mere three percent of its land surface – than any other continent. Both the Ethiopian and Australian regions have ten times as much desert. Fourthly, a huge part of the continent is blessed with over 60 inches of rain per year, giving rise to the world's most extensive rainforest and, with the Amazon, the world's greatest river. Over 30 percent of its surface is classified as rain-forest with a further 22 percent being woodland or open forest. Despite the great rainforests of the Congo basin, the equivalent figure for the Ethiopian region is only nine percent. The combination of a huge mountain chain, extending nearly halfway around the world along a north to south axis, with vast rainforests centered along the equator has produced unique opportunities for birds to speciate in glorious isolation.

One of the best examples of these factors at work can be found by making a journey from the Pacific coast of Ecuador to the high Andes, from Guayaquil to Quito. Though only 250 miles as the Condor soars, changes along this route are dramatic, heightened by the fact that one is virtually on the equator the whole way. Most dramatic of all is the relatively short journey through the cloud forest that covers the Pacific slope of the Andes between 5,000 and 10,000 feet. Here a road zigzags Alpine-style through the forest and produces new species at virtually every zig and zag. In an area lacking seasonality, where day length and climate barely vary throughout the year, a relatively small difference in altitude can produce new habitat and new opportunities for birds. With trees and other vegetation flowering and fruiting throughout the year, birds can afford to become both highly specialized and completely resident. And that means they can develop as species within a surprisingly short distance, one from another. For the bird-watcher this journey is tremendously exciting, for the ornithologist it is nothing short of miraculous.

The proliferation of species is most obvious in the small perching birds, members of the huge order of Passeriformes. Indeed, the sub-order of Tyranni, which is virtually confined to the Neotropical region, contains over 12 percent of all the world's bird species. Outstanding numbers of species are found, for example, among the hummingbirds with 319 distinct species, all but a handful of which are found only in South America. Also, there are 366 tyrant flycatchers, 221 antbirds, 212 ovenbirds, 90 contingas, 59 manakins and 47 woodhewers which are either totally or all but confined to the region.

With such a wealth of species at family level it is perhaps not surprising that the visiting bird enthusiast is often confused and daunted by the problems of identification. Indeed, one American ornithologist advises all of his friends to ignore the tyrant flycatchers completely during their first visit to South America. While this is almost impossible to do, a glance at one of the better field guides to any South American country is sufficient to present such a viewpoint in a particularly graphic manner. Illustrations depicting 40 or 50 more or less identical tyrant flycatchers are formidable, but so are those that show 25 woodcreepers or a similar number of antbirds. These birds are so alike that the proliferation of hummingbirds seems easy in comparison – which it is not.

Even among the generally more mobile and thus widespread larger birds, South America scores highly in terms of endemism. The rheas that roam the open grasslands of the southern part of the continent are found nowhere else, though they do have ratite relatives elsewhere in the southern hemisphere. While the Greater Rhea is a bird of the pampas, the Lesser Rhea inhabits the puna zone of the Andes at considerable altitude. Similarly, the tinamous, of which there are some 46 distinct species, are a group of ground-dwelling birds that are essentially Neotropical in distribution. Though they are often known locally as "partridges" and bear a strong resemblance to those birds, they are much closer to the rheas in origin and comparison with the gamebirds of the Old World is a straightforward example of convergent evolution.

Female tinamous are not only larger than their males, but are generally more boldly colored. As with several other groups of bird, where the female is more colorful, it is the male that incubates the eggs and cares for the young. Such an arrangement, called role reversal, is found in the phalaropes and among several other groups of wading birds. When incubating, the male relies totally on his highly camouflaged plumage to

avoid detection and may, therefore, allow a very close approach. Certainly, there are plentiful examples of males allowing themselves to be touched or even picked up from their nest. These ground-dwelling birds are experts at avoiding detection by standing stock-still and by putting whatever cover is available between themselves and an intruder. Some live among open grasslands, but many are forest birds and notoriously difficult to see. As they make good eating, such a strategy is essential to their survival. They are poor fliers and frequently injure themselves when flushed among trees.

The two species of screamer are aptly named. These are large water birds that are usually regarded as relatives of the ducks, geese and swans. They have long legs and large, unwebbed feet, but spend much of their time among marshes. They have large wings and often soar like storks on rising

The Hoatzin (right) is a bird of riverside trees and bushes where it often forms loose feeding flocks. Its long tail and broadly rounded wings are ideally suited for gliding from tree to tree, but are quite inadequate for sustained flight. Soon after hatching (opposite) the Hoatzin chick is capable of clambering about among tree branches aided by strong legs and feet and a claw at the bend of the wing. It is even capable of swimming should it fall into the water, a skill soon lost when it develops proper feathers.

warm air. Another waterside bird that is also confined to the Neotropical region is the remarkable Hoatzin. This large, chunky, golden-colored bird is placed in a family by itself. Like the screamers it is equipped with a large spur at the bend of the wing. But while the screamers use their spurs as weapons, the Hoatzin uses its spurs as a "hand" to help it climb among riverside bushes and trees. The Hoatzin is a very strange bird indeed for, like the ancestral bird *Archaeopteryx*, it is a poor flier and uses its wings mainly for gliding from one tree to another. Its nest is built over water and the young start clambering about long before they are feathered. If, by accident, they fall into the river or stream, they are capable of swimming, a skill that is lost in adulthood.

The three species of trumpeters and the monotypic Sunbittern also form totally Neotropical families, as do the potoos, seed snipe and oilbird. The seed snipe are a strange group that are generally regarded as being closely related to the waders. They are chunky and thickset in build with short legs

and a small, seed-eating bill. Like snipe, their flight is fast and erratic and, also like those birds, they are well camouflaged. They frequent open habitats from the high pastures of the Andes to barren sea coasts. One species, the White-bellied Seed Snipe, is found only in the sub-Antarctic parts of southern Argentina and remote Tierra del Fuego, though it also nests on the Falkland Islands. Like the other three species, it lays a wader-like clutch of four conical-shaped eggs.

The potoos, of which there are five species, are similar to nightjars in both appearance and habits. They are long-winged and long-tailed, marked with highly cryptic camouflaged plumage and with soft margins to the flight feathers for silent flight. Like the nightjars, they are nocturnal insect hunters and spend the day hidden away and inactive. Unlike the nightjars, potoos roost and nest in trees rather than on the ground. They lay a single white egg in a tree crevice and incubate in a vertical position.

The most remarkable member of the order Caprimulgiformes, or

Three distinctive, but unrelated, families of birds that, along with several others, are to be found only in the Neotropical region. The Lesser Potoo (right) is a nocturnal, nightjar-like bird, that spends its days pretending (successfully) to be a broken branch. Like other such birds, it has a repetitive but characteristic call. The Sunbittern (above opposite) is frequently kept in captivity. It is a bird of waterside margins that boasts an elaborate display. The Oilbird (below opposite) is a remarkable nocturnal fruit-eating species that lives colonially in the total darkness of caves where it finds its way around by a bat-like echo-location system.

nightjars, is the Oilbird. Strangely gingery-brown in color and marked only by very large eyes, this is a nightjar that has taken to a diet of fruit and, as a result, has developed a hooked, parrot-like bill. It has also become a cave-dweller and, like the bats, has developed an echo-location system to help find its way in complete darkness. Like the fruit-eating bats, the Oilbird is gregarious, forming nesting colonies and communal roosts. The breeding routine is slow with over a month of incubation and over 100 days of growth prior to fledging. After a couple of months of rich fruit-feeding the young become very fat and weigh half as much again as their parents. At this stage and still covered in white down, they were formerly collected by local people and boiled down to produce a useful and pleasant, nonsmelly oil.

Seeing nightjars anywhere in the world is always difficult and usually dependent on locating the birds by their repetitive calls. In the case of the Oilbird, viewing is almost always at a nesting colony, which may pose a problem of disturbance. At a well-known colony in Trinidad visiting had to be restricted because of the effects of too many tourists disturbing the birds.

The Sunbittern is a waterside bird mostly seen wading along the margins of forest streams. Mottled in browns and blacks, it is a well-camouflaged

Although a few species have managed to colonize northward into the Nearctic, the vast majority of the world's 315-odd hummingbirds are endemic to the Neotropical region. Nowhere, perhaps, are there more species than among the coastal cloud forests of the western Andes from Colombia to Bolivia, where every change of altitude brings different flowering plants and different hummers to pollinate them. The Sword-billed Hummingbird (above) boasts the avian world's largest bill in proportion to body size, a unique adaption to a single flowering plant. The Red-tailed Comet (right) is found at high altitude in the temperate zone of the northern Andes.

The Copper-rumped
Hummingbird (above) is a
member of the large genus
Amazilia. It ranges from
Venezuela to Trinidad and
Tobago. The Violet-tailed Sylph
(below) is found from Colombia
to Ecuador and is sometimes
considered no more than a
subspecies of the Long-tailed
Sylph which, while being found
from Colombia to Northern Peru,
has a more easterly distribution.

bird that can turn itself into an avian jewel by spreading its wings and tail like a peacock in display. In its forested habitat it more closely resembles an Australian lyrebird though, unlike that bird, it is mostly silent.

As we have seen, however, when we come to the smaller birds the Neotropical region is truly in a class by itself. Perhaps no group shows the variety of species within a particular family better than the hummingbirds. Though they vary in size from that of a thrush to the world's smallest bird, at two and a half inches in length, most are about three and a half inches. They are generally brightly colored, with a vivid emerald green broken by patches of bold irridescence that change in hue with the angle of the light. These birds are nectar-feeders, with shape and size of bill adapted to a huge range of blossoms, but they also take insects. They have relatively long wings, with curiously square-tipped feathers forming a solid fan that creates power on both the upward and downward strokes. In fact, they are the only birds in the world that can actually fly backwards. All but a tiny handful of the 319 species are found exclusively in the Neotropical region.

The world headquarters of the hummingbird family lies astride the equator in the tropical rainforests of the Amazon and into the high Andes and Pacific cloud forests of Ecuador. In either direction, north and south, the number of species falls off rapidly. One species, the Rufous Hummingbird, migrates the length of the continent to spend its summers as far north as Alaska.

These are fast-flying birds that, as their name implies, produce a loud hum when their wings are beaten. They are solitary and highly territorial, flitting swiftly from one feeding plant to the next. Such high mobility makes them difficult to observe and, together with their general similarity one to another, even more difficult to identify. Although each species has a particular bill structure to enable it to feed from a particular flower or group of flowers, these differences are often subtle and of little use for identification. Some have tiny tube-like bills, while others have curiously decurved, sickle-shaped ones. Quite outstanding is the Sword-billed Hummingbird, which has a tube of a bill that is actually longer than the whole of its head, body and tail. It was from observation of this bird that a botanist was able to predict the existence of an unknown flower, and later to search for and find it.

Hummingbirds also vary considerably in tail shape, though the more flamboyant examples are totally to do with display rather than ecology or flight efficiency. The Booted Raquet-tail, for example, has two extended central tail feathers that end in round disks; the Crimson Topaz has the rufous outertail feathers extended and bent inwards to cross each other; the Long-tailed Sylph has a remarkable blue forked tail, and so on.

Although hummingbirds are totally confined to the New World, they do have their Old World equivalent in the sunbirds of Africa and Asia. These birds, too, feed on nectar and are brightly colored with irridescent plumage. There, however, the similarity ends, for while hummingbirds feed on the wing, sunbirds feed while perched, leaning acrobatically to insert their bills deep into flowers. Once again we have an example of birds evolving quite

The similarity in appearance between the Old World bee-eaters and the New World motmots is well known in these two photographs. The Green Bee-eater (left) is a widespread resident of the Oriental region, while the Blue-crowned Motmot (below) ranges from Mexico to Argentina. Both groups are insect eaters and both excavate long tunnels in earthern banks as nest sites.

Though such a characteristic element of the South American avifauna, the macaws are always elusive among the high canopy of the forests they inhabit. These are long-tailed, long-winged, fast flying birds that seldom pause in the open to give good views. Sometimes large numbers will gather at a particular spot along a river bank to eat earth rich in minerals. They are, however, widely kept in captivity, especially the Blue and Yellow Macaw (above opposite) that ranges from Panama to Brazil and the Scarlet Macaw (below opposite), found from Mexico to Brazil. The Hyacinth Macaw (left), in contrast, is found only in central Brazil in the forests of the Amazon basin.

separately to take advantage of a particular source of food, and coming to resemble one another. The present example of this convergent evolution has, however, stopped at the flight level because each group manages to feed successfully in its different way.

Similar forces have been at work in making the Neotropical toucans resemble Old World hornbills in form and behavior, and the motmots of America structurally alike to the Afro-Asian bee-eaters. The motmots are a colorful group of upright-perching, insect-eating birds that, like the bee-eaters, beat their prey before consuming it. They also excavate tunnels in sandy banks and lay their eggs in a chamber at the end. They are, however, more closely related to the kingfishers, a family that is very poorly represented in South America and clearly originated elsewhere.

One of the most characteristic groups of South American birds are the parrots, yet informed opinion places these highly colorful birds as having their origins elsewhere and the diversity of species as an example of radiant evolution. Nevertheless, the huge macaws that fly so fast over the tree tops of the Amazon jungles are unique and can be found nowhere else. There are 15 distinct species including the Blue-and-yellow Macaw, which is widely kept in captivity. Not surprisingly, this was one of the first South American birds to be imported into Europe by the conquistadores. It is among the most successful of the parrots with a range that covers much of tropical South America. In sharp contrast, the smaller Spix's Macaw was regarded as extinct in the wild until the discovery of a single individual in eastern Brazil in the late 1980s. About 15 are in captivity and this last individual was

Most beautiful and elusive of all the world's trogons, the Resplendent Quetzal is found only among the forests of Central America from southern Mexico to Panama, though its close relatives extend southward as far as Bolivia and northern Brazil. Its extended tail feathers have always been prized by man.

seized by unscrupulous traders in 1991. It may well be that this bird was the last of its kind and the species is destined for the extinction list.

Equally at home among the tropical rainforests are the trogons. This family, well represented in the Neotropical region, has always caused zoogeographers immense problems. The 35 species are widely distributed throughout the tropics, being widespread in both Asia and South America. The family is divided among nine distinct genera, with one dominant genus in each of these continents. The genus *Harpactes* is found only in the Oriental region, the genus *Trogon* only in the Neotropical. Other smaller genera occur in Africa and in South America. The species radiation from a single ancestor in South America and Asia does little to help us understand the origins of the family as a whole. South America does have more distinct

Although trogons are found in many parts of the tropics, there are more species in South and Central America than anywhere. Whether the birds are a particular ancient line, or whether, as seems unlikely, they have colonized new continents, it is impossible to say. The Violaceous Trogon ranges from Mexico to Brazil and has even managed to colonize Trinidad. Did it fly? Or was it a colonist before Trinidad became an island?

genera, but one should not infer that this continent is therefore the original home of all the trogons.

Trogons are medium-sized birds marked by a characteristically thick-set, chunky appearance and a strong head and neck, with a deep stubby bill. They have short, rounded wings and long tails, but are strong fliers. Asiatic species feed flycatcher-fashion on insects caught in the air. Neotropical birds catch flies, but also take large quantities of fruit. In general, these are colorful forest birds, but their habit of sitting stock-still for long periods, even when closely approached, makes them difficult to observe and easy to overlook. Like many other arboreal groups, including the woodpeckers, they have two toes pointing forward and two back. Again like the woodpeckers, they excavate their own tree nest holes, but they are only able to do so in the decaying wood of dead trees. One species, the Violaceous Trogon, actually prefers to excavate its nest within the paper nest of a social wasp. As it digs it consumes both insects and grubs and seems to be immune to stings.

Outstanding among the trogons is the Resplendent Quetzal, a strong contender for the title of the world's most beautiful bird. This vividly green bird has the uppertail coverts extended to form a "tail" that is at least twice as long as the bird itself. It occurs only in Central America, where it is difficult to locate, and was regarded as sacred by the Mayas. Even with its extraordinary tail it still nests in holes, a form of behavior which quickly destroys its finery during the breeding season.

The rich seas of South America have produced some of the greatest seabird colonies in the world. These King and Guanay Cormorants have built earthen mounds as nests as Punta Tombo, Argentina.

The migration route of both Eskimo Curlew and Lesser Golden Plover from their Canadian breeding grounds to the South American pampas seems eccentric on a normal map projection (left). But on the projection (right), where great circle routes emerge as straight lines, the Atlantic route is shown to be the shorter.

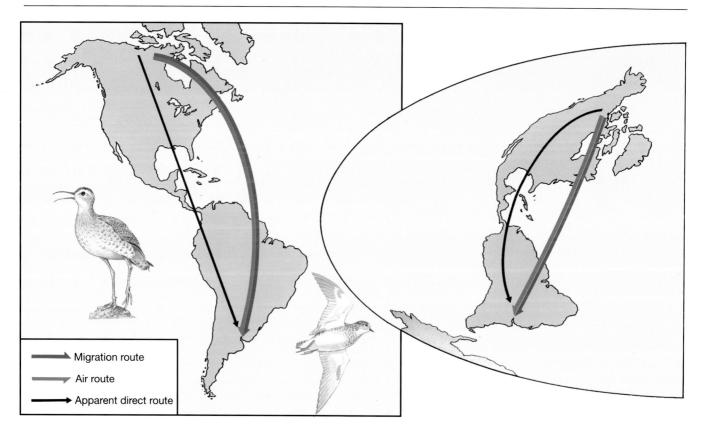

Migration route

Air route

Apparent direct route

Given its remarkable depth in terms of endemic families, genera and species, the Neotropical avifauna remains surprisingly self-contained. Only a few species managed to cross the Central American gap during the period when North and South America were separated by narrow seas. Most of the birds that have spread northwards have done so since the land bridge has been established, but numbers remain small. Movement in the reverse direction, from north to south, has been more extensive. The cardinals and tanagers, for example, that originated in the Nearctic have colonized southwards and developed into a wide range of distinct species. In fact the tanagers are now an integral part of the South American avifauna.

Not all of this movement from north to south is, however, colonizing. Every autumn huge numbers of Nearctic birds fly southwards to winter quarters in the Neotropical region. Wildfowl, waders, herons and millions of warblers and flycatchers are involved, movements that find a parallel in the bird migration systems of Europe–Africa and Siberia–Asia. As with migrants elsewhere, different species have adopted distinct migrational tactics. Many of the Nearctic waders, for example, fly virtually nonstop from the far north to their wintering grounds as far south as Argentina. For them speed is the tactic, rather than refuelling on the way.

The classic case in this respect is the Eskimo Curlew, now all but extinct. In the past, these birds left their breeding grounds in the northwestern Canadian tundra to gather in enormous numbers in Labrador in the autumn. There they would feed voraciously before setting out on a huge loop of a flight over the Atlantic ocean to make a landfall among the marshes of northern Argentina. If such a route seems illogical, a glance at a map of the world air routes, in which great circle routes appear as straight lines, is sufficient to show that these birds actually took the shortest route between two points. Strangely, the Eskimo Curlew returned northwards via a completely different route that, in the Nearctic at least, followed the mid-continent along the Mississippi flyway. Just why these birds should take a more leisurely route in this direction remains a mystery, though other waders here and elsewhere follow similar loop migrations.

Huge numbers of other birds also make a sea crossing when moving both southward and northward, even though they could follow the land bridge through Panama in both directions. Such oversea routes could have evolved during the period when North and South America were separated by sea. But evolution via adaption can and does work remarkably quickly, and there must be a clear advantage in crossing the sea rather than following the land or the process would have died out. Again a glance at the air routes map will show the advantages of crossing the Atlantic and the Gulf of Mexico. Landfalls of birds that have spent the winter months in the Neotropical region occur all along the eastern seaboard of the United States, but the Gulf coast of Texas is an outstanding migration "fall out" station in spring. Even species that breed well to the west follow a line along the Mississippi rather than a more direct route. Thus both Mourning and Connecticut Warblers, for instance, are virtually unknown in the southwestern United States in spring.

The Puerto Rican Tody is a member of an exclusively Caribbean bird family consisting of only five species. Generally regarded as being most clearly related to the motmots, these are dainty little birds that hunt flycatcher-fashion from an exposed perch. A rattling noise in flight is produced by the outer primaries fluttering through the air.

Migration is, of course, a rather hazardous business and a huge number of birds get lost every year. Some doubtless manage to reorient and find their "correct" home, but others seem destined to die. It was, perhaps, a migratory flock, or perhaps several individuals, that was responsible for one of the most remarkable avian colonizations in recorded history. Initially, Cattle Egrets arrived in South America during the first half of the present century. They were soon established in Guyana and rapidly spread along coasts and rivers to occupy virtually the whole of the New World from the United States–Canadian border southwards to Argentina. That this same remarkable bird has also managed to colonize Australia during the same period shows clearly that there are vacancies for a "new" species in many parts of the world. The Cattle Egret is named for its association with herds of domestic stock, alongside which it feeds on insects disturbed by these large mammals, and it will associate with other large mammals wherever they occur. Such an association emancipates the birds from a dependency on marshlands and opens up completely dry areas for colonization. Although they usually nest colonially among waterside trees, they will also do so in other areas such as coastal cliffs. It is a short move from feeding alongside a large animal to following the plow and, in many areas, "tractor egret" would be a more appropriate name. Similar, though more ancient, colonizations have been used to explain the presence of the Southern Pochard and the whistling ducks in the Neotropical region.

Though essentially part of the region, the West Indies have a rather poor avifauna that probably results from overseas colonization from tropical

Above: *Frigatebirds are found at many places around the coasts of tropical South America, as well as elsewhere in the world. The huge inflatable sac on the neck of the male is wobbled alluringly whenever a female appears – and it seems to work.*

Left: *The most recent case of a classic colonization occurred when Cattle Egrets flew from Africa to South America.*

Two colonists of the Galapagos Islands, Ecuador arrived from quite different origins. The Galapagos Flycatcher (right) is a member of a genus that is widely spread in South America from the West Indies to Bolivia and Brazil. The Waved Albatross (opposite), in contrast, belongs to the genus that includes all of the world's large albatrosses based in the North and South Pacific.

South America. The birds of this subregion include mockingbirds, vireos, tanagers, tyrant flycatchers and others that are more appropriately regarded as Neotropical in origin. The West Indies, however, do have two endemic families – the Palm Chat and the todies.

The Palm Chat is abundant and widespread throughout the island of Hispaniola, but is found nowhere else. It is unusual in building a huge communal nest among the bases of palm fronds, within which each pair creates a separate chamber to lay their eggs in and rear their young. Some authors consider these birds as close relatives of the waxwings, but most prefer to regard them as a monotypic family.

There are five species of tody, all confined to the islands of the West Indies. Four species differ only in the color of breast and flanks and by occurring on separate islands. It would therefore be easy to regard them all as no more than subspecies, and, in any case, they certainly form a superspecies and share a common ancestor. The fifth species is found only on the island of Hispaniola and is physically distinct from the others by virtue of its narrow bill. These are tiny, green-backed birds, totally insectivorous, that catch their prey on the wing flycatcher-fashion among shady forest glades. Like so many other forest insect-eaters they are easily overlooked and difficult to observe. They may be related to the motmots or the kingfishers, for like both these species they excavate a nesting tunnel in a soft earthen bank and their eggs are white.

Of all the Neotropical birds, none have been the subject of so much speculation and study as those of the Galapagos Islands. Lying only 600 miles west of Ecuador, these are volcanic islands that rose lifeless from beneath the Pacific ocean. For this reason they are home only to species that have colonized from elsewhere or, more accurately, to species whose ancestors were colonists. It is likely that these remarkable islands – the ornithology of which began with the visit of Charles Darwin in 1835 – were first colonized by seabirds that used them solely as resting and later nesting places. Such birds take and need nothing from the land and can exist even if the islands are bare and barren. Later, following colonization by plants, there were opportunities for vegetarian birds to glean a living, and later still, following colonization by insects, a window of opportunity opened for birds dependent on animate food.

The Andean Condor survives among the mountains and coastal deserts of western South America. Its nearest living relative, the California Condor survives only in captivity in the United States. Will one follow the other into ultimate extinction at the hands of man?

It is possible for each of the various niches that develop on islands like the Galapagos to be filled by an appropriate colonist from elsewhere. Such a scenario is, however, unlikely and has not occurred here. What did happen was that the islands were colonized by a finch from mainland South America about a million years ago. This original finch colonist, which has since become extinct elsewhere, spread through the widely scattered archipelago and developed different tactics to deal with the circumstances in which it found itself. Over the years, different island populations became ecologically distinct and when individuals (or flocks) moved to other islands within the group they were able to coexist alongside the native species that had evolved separately.

From the original ancestor no less than 13 distinct species have evolved. The Geospizinae are today called Darwin's finches. Only a handful of other South American land birds have managed to colonize the Galapagos – the Yellow Warbler, two tyrant flycatchers and a mockingbird that has managed to radiate to form four distinct species. Otherwise, the only birds are seabirds; yet even here the Galapagos are unique in boasting the only albatross to breed on the equator and the only penguin to penetrate the northern hemisphere. As a postscript to the Galapagos story, another "Darwin's finch" was found on the remote Cocos Islands in 1891.

Throughout this account of the incredibly rich Neotropical region we have stressed that the uniqueness of its avifauna is based mainly on the

The Purple, sometimes called
Yellow-legged Honeycreeper is a
member of a Neotropical group of
nectar-eating birds that are mostly
highly colored as are so many of
the birds of this, the richest
avifauna in the world.

extent of dense rainforests, centered around the tropics, where birds have
little need of movement. We have seen that even small changes in altitude
can be sufficient to provide the isolating mechanism that enables species to
develop. Yet when the Andes and the Rockies are treated as one chain these
same mountains extend over almost the whole of the Americas. Following
this highway a number of species have moved southwards into the region.
They include pipits, the Short-eared Owl and even the Shorelark (Horned
Lark in North America) – all high latitude or high altitude species of
northern origin. These same mountains are also home to the huge Andean
Condor, a scavenging vulture that is today the only wild bird in the world
to bear that name. (Until the 1980s a few Californian Condors still managed
to eke out an existence in the hills above Los Angeles, but numbers were
small and productivity incredibly low. The remaining birds were then
captured to form the basis of a captive breeding scheme. As this is being
written some are about to be released into the wild.) The Andean Condor is
fortunately still alive and well in good numbers. It inhabits the mountain
strongholds of the high Andes from Colombia to Tierra del Fuego, though
it also descends to sea level along the wildest coastlines. Even though it is
not yet endangered, the Condor faces serious threats and is persecuted by
peasants who are not overfond of losing their domestic animals to this great
bird. Protection is essential, but protective legislation in South America is
barely worth the paper it is printed on.

CHAPTER NINE
SURVIVAL
THE FUTURE OF BIRDS

Like most other North American duck, the Redhead has suffered a steady decline in numbers due almost entirely to overhunting. The frontiersman spirit, combined with the right to carry arms, has decimated much of the continent's wildlife in little more than a century and a half.

Hunted to the brink of extinction, the recent more enlightened attitude towards birds of prey in Britain has enabled the Red Kite to regain some of its numbers, although, as elsewhere in Europe, it seems unlikely that it will ever be as abundant as it was 200 years ago.

There is, it seems, little doubt that our planet will increasingly become a single-species space capsule. The human population will continue to expand at the most alarming rate, making ever greater demands on the earth's resources. Fossil fuels that, together with medical advances, are largely responsible for this increasing tide of humanity are finite and will eventually run out. We may – or may not find substitute sources of power, though rediscovering the windmill and watermill are not the answers. Along with this ever-increasing population there will be a parallel demand for ever-increasing production and ever-rising standards of living. Such demands by larger and larger numbers of people will mean more houses, more cars, more roads, more airports, more hotels and resorts, more intensive agriculture, more offices and more exploration for more and more remote raw materials. Depressing, isn't it?

There is also little doubt that although nations would like to do something to control their populations, they will fail to do so. China, the only nation so far to face up seriously to the overpopulation problem, has failed. India, by quite different means, has also failed to halt its incredible population boom. Perhaps some natural or human-initiated disaster will strike before it is too late? Could there be further wars as the most powerful nations seek to ensure more than their fair share of oil? Could the nuclear holocaust actually happen? Could pestilence in the form of AIDS, or some such, wipe out huge numbers of people worldwide? At present the signs are not likely.

Thus we have and will have a planet that is totally dedicated to the needs, desires and whims of a single, highly aggressive species. Man will turn his attention increasingly to the last great wildernesses – to Antarctica and to the great rainforests. He will seek to control rivers, to exploit deltas, to drain wetlands and flood valleys. No part of the planet will escape, except those areas that man himself designates "natural" for his own enjoyment or from a rush of conscience. Taking the long-term view will become increasingly more impossible, as it already is for publicly quoted companies dependent for their very existence on short-term profits. There will be a

The chances of anyone seeing a Californian Condor soaring over its Rocky Mountain home ever again depend entirely on the success of a last ditch captive breeding program in the United States. That such a magnificent bird can come to such dire straits is a poignant commentary on man's relationship with the other inhabitants of this planet.

few voices in the wilderness arguing the case for a reservoir of nature in case it should prove useful in the future, but such voices will be submerged, as they always have been, by the mass of self-seekers. Even more depressing?

It is against this background that we have to examine the future of wildlife and the future of birds in particular. The birds of Europe and North America are among the best protected and most studied in the world and an examination of their status, as we come to the end of the twentieth century, may give us some idea as to what birds elsewhere in the world can expect.

The birds of Europe and North America have to live alongside humans – some can, others cannot. The House Sparrow, Starling and Herring and other gulls have done very well and prospered. They are abundant,

204

widespread and increasing. In contrast the Great Bustard and White-tailed Eagle have been exterminated in Britain, though the Eagle has been successfully reintroduced. There too the Osprey, Avocet, Black-tailed Godwit and Savi's Warbler were all virtually exterminated, but have managed to reestablish themselves naturally, while others such as the Red Kite, Marsh and Montagu's Harriers and Dartford Warbler totter along at the edge of disaster and survive only because of the efforts of a few British conservationists. These are the "star stories" of birds that have disappeared or seem likely to disappear as Britons set about changing the face of their country to suit their own needs. But what of the others?

In North America the Passenger Pigeon, Carolina Parakeet and Labrador Duck were all exterminated during the nineteenth century. The California Condor made it into the twentieth century and may, via a captive breeding program, scrape into the twenty-first. Similarly the Eskimo Curlew, once thought extinct, still apparently manages to survive. Of course, the successes of the Whooping Crane breeding project should be celebrated, but its survival still rests on a knife edge. It is as easy in North America as it is in Europe to talk of conservation successes, but is it really a success to save the last few individuals when we (as a species) have exterminated all the rest? The story of the distribution of the Greater Prairie-chicken is paralleled in Europe by that of the Great Bustard. The plight of the Least Tern is the same as that of the Little Tern, while the decline of the Roseate Tern is the same in both continents.

In America, as in Europe, there are successful species, but the booming population of gulls, Fulmars, Starlings and English Sparrows cannot be regarded as adequate compensation for the demise of some of North America's most spectacular birds. Overall the population of North American ducks continues to decline due to overhunting. But what can we do to reverse the trend? In Europe virtually every species of bird of prey is in decline. There seems little that can be done there too!

The rule for survival is relatively clear. If it can live alongside man a species will survive. If it can exploit man's world then it will prosper. If it can do neither, it will either disappear or it will survive due to man's benificence. So at one end of the response chain we have the gulls, in the middle the flycatchers, and at the other end, the birds of prey.

Of those species that actually exploit man's world, the gulls are probably the best example. During the past fifty or sixty years these basically coastal birds have been able to move inland during the winter only because man has provided the perfect conditions. The construction of large lowland reservoirs near large centers of population now offers them secure roosts where formerly there were none. The roosts in turn provide what are essentially scavengers with the opportunity to exploit the food that we throw away. So a combination of rubbish tips and reservoirs has created a sort of gull paradise and numbers have boomed as a result.

Such a population explosion inevitably has side effects. Gulls are large, gregarious birds that pose a serious threat to the modern jet engine. Although accidents involving death are thankfully rare, it can only be a

THE SURVIVAL WORLD OF BIRDS

matter of time before a fully laden airliner crashes on takeoff killing a large proportion of its passengers and crews. While taking due account of the possibilities of bird strikes, planners are bound to allow airports to be developed near both reservoirs and rubbish tips, for all three are inevitably located near centers of population.

Gulls are also highly likely to be infected with diseases such as botulism by virtue of feeding among the filth of human waste. Strangely enough this does not seem to worry officials of the water supply companies on whose reservoirs they roost in their tens of thousands. Yet four out of five people infected with botulism will die if not quickly diagnosed and treated, and four out of five of those treated will also die. So far, once again, there has been no case of a mass outbreak of botulism due to gulls but is this, too, only a matter of time?

In the same way that a human population can get out of hand, so too can a bird population. The gulls that spend the winter inland around our cities may spend the summer at the network of bird reserves that extends around our coasts. There, the sheer numbers of a single species may pose a threat to scarcer species that are the major object of conservation efforts. To see eggs and chicks of Avocets, for example, being taken by gulls is a heartbreak to any reserve warden. So, not surprisingly, efforts are made to control gull numbers by nest raking or taking the eggs. Bird conservation is in part a matter of choosing between species.

Not all species that have flourished by exploiting man's activities pose problems as a result. The Fulmar, once confined to a few isolated rocky outcrops, has increased and spread to many coastline areas. This growth is, if anything, even more spectacular than that of the gulls, for these are long-lived birds that do not breed until they are seven years old and then only lay a single egg each year. Gulls, in contrast, breed at two to four years old, lay three or four eggs and therefore have a much greater potential to expand their numbers.

The spread of the Fulmar, and incidentally also of the Kittiwake, has been ascribed to their ability to feed on the huge amounts of fish waste produced by modern deep-sea trawling. As fish stocks around the coasts continue to dwindle due to uncontrolled overfishing, we should expect a reversal of the remarkable Fulmar boom, though as yet there is no sign of this occuring.

Gulls, Fulmars, House Sparrows, Starlings and so on may be the most obvious species to have exploited man's activities, but they are far from being the only ones. Where, one could reasonably ask, did Swallows (Barn Swallows in North America) nest before man built such admirable sites for them? Well, Swallows nested in caves, which are in very short supply in inland Britain. Swifts were also cliff nesters, while Sand Martins not only needed cliffs, but also particularly soft ones in which to excavate their nest holes. The growth of the construction industry has created a network of sandpits throughout the lowlands to the delight of the Martins. There seems little doubt that all of these aerial species have benefited from man's activities.

Other species may be less directly linked to man, but have nevertheless

Two Whooping Cranes at their traditional wintering grounds at Aransas Wildlife Refuge on the Texas Gulf Coast. The recently successful captive breeding program to save this great bird should not blind us to the predicament of other, less spectacular species that similarly totter on the verge of extinction.

The aptly named Wood Duck has declined in numbers for two quite distinct reasons. Like most other North American ducks it has been overhunted, but equally important has been the destruction of the lakeside woodlands in which it lives.

benefited from the changes he has brought about in the countryside. It is beyond the scope of this book to examine the financial and fiscal advantages of planting trees or, come to that, the policies of government that encourage such activities. It is the results that concern us, though a change in the tax laws could see wholesale felling over large areas of Britain. The new lowland conifer plantations, so often regarded as birdless by ornithologists, have in fact created a safe haven that a booming population of Woodpigeons finds perfectly suited to its nesting and roosting needs. Yet these same birds are persecuted without respite by farmers whose crops they damage. Goldcrest, Coal Tit and Great Tit have all prospered due to the provision of these new habitats, and both Common Crossbill and Firecrest have colonized parts of southern England for the first time.

We have already mentioned classic cases of decline and destruction that form the other side of the coin, but the case of the Red Kite in many ways encapsulates the problems of living with and alongside man. In Shakespeare's day these large, fork-tailed birds of prey were a common sight throughout Britain. In London they even took food from the hands of children. A gradual clean-up of the cities during the nineteenth century was bound to cause a decline, but this was accompanied by one of the great anti-bird campaigns in which the Red Kite had a price on its head. Details of this

bounty system are preserved in parish records throughout the country, though just why these birds were singled out for such treatment is difficult to understand. The result, however, was outstandingly successful. Kites were exterminated from one county after another until only a handful remained in central Wales. There a band of dedicated conservationists battled to save the species against local prejudice, vandals and egg collectors. After almost a hundred years there are more Kites in Wales than ever before, but their numbers are still precariously low and they are still poisoned and robbed of eggs every year.

Although not subject to the same degree of deliberate persecution and in nowhere near such a precarious state, the population of Little Terns gives more than passing cause for concern. Here, the problem is of a bird trying to live alongside man and regularly failing to do so. Little Terns nest on beaches just above the high-tide level, and as long as beaches were the haunt of a few beachcombers and fishermen the Terns had this habitat to themselves. With the coming of the railways and, more particularly, cars the beaches became the focus of holidaymakers and trippers and the Little Terns lost nest after nest. Traditional colonies were quickly decimated and only those that were remote, or of awkward access, remained viable. Fortunately, the British are intrinsically lazy and loath to walk further than

Habitat creation may, even inadvertently, benefit species such as the Common Crossbill, which has expanded its range as a result of conifer planting in lowland Europe.

The rapid decline in the numbers of Little Terns is directly correlated to the increase in human leisure and mobility. Only specially protected beaches can accommodate both Little Terns and holidaymakers, otherwise the birds are forced to use only the most inaccessible shingle coastlines.

they need, so beaches that remain remote may still form a home to these reduced colonies. Unfortunately, it is now the very people who support conservation, people who are prepared to make the effort to get to an unspoiled coastline, who are causing remaining colonies to desert their nests.

Today, wardens entrusted with the conservation of a Little Tern colony do everything they can to persuade the birds to nest inland away from the beaches as soon as they arrive and well before the arrival of visitors. Sometimes they succeed only for another, perhaps natural, disaster to strike. At one colony all of the warden's work has been wiped out over the years by a succession of Kestrels, foxes and egg collectors.

Britain and America's birds are in good hands and the conservation movement is now a force to be reckoned with. Nevertheless, pressure on the land in the ever-increasing search for a higher standard of living – more cars, holidays, golf courses, and so on – poses problems that simply protecting species does not cover. What is the point in guarding nests, imposing penalties for disturbance, setting up nature reserves and taking other measures, if birds are deprived of their natural feeding grounds?

Wildfowl, ducks, geese and swans, together with the waders, are a particular case in point. Estuaries are among the most important feeding grounds for these species and support hundreds of thousands of birds, both on migration and in winter. Yet industry sees estuaries as wastelands, sites

Arguably the world's most beautiful tern, but also among the most endangered, the Roseate Tern has suffered a severe decline in numbers in many parts of its range. Even well protected colonies are not immune and there seems little that conservationists can do to halt the process.

where power stations, docks, refineries and factories can be built without upsetting too many people. They are reclaimed for agriculture and used as dumping grounds for waste, for the same reasons. As a result, a large proportion of estuaries are endangered.

Everyone knows that we already produce too much food and too many cars, that power stations produce acid rain when they burn fossil fuels and that dumping waste pollutes the sea and beaches. But it appears that someone has a scheme for using every tiny scrap of land, no matter how unlikely it seems. We drain it, flood it, plant it, fell it, extract it, use it for infill and, of course, build on it. The alternative is to conserve it – but there is no money in that!

It is comparatively easy to take the experience of Europe and North America and apply the lessons elsewhere. In this way we can predict that as more and more countries aspire to the sort of living standards westerners enjoy, their birds will be affected by the same forces. The world's resources will be exploited as long as it is economically viable to do so.

The ability to live alongside man has enabled populations of Herring Gulls to explode in many parts of their range. Their exploitation of rubbish tips, coupled with secure roosts at newly constructed reservoirs, have been the major factors in the success of this large and powerful bird.

The rainforests, for example, will be felled until the value of the timber they produce is less than the costs of extracting and exporting it. Then the remnants will be regarded either as trendy nature reserves, a sop to the world conservation movement, or as home to an increasing population of impoverished subsistance farmers. Rainforests are complex ecological systems that are perfectly suited to the land and climate in which they grow. Once these forests are felled, the rain, which is so crucial to their original creation and maintenance, has a very adverse effect on the soil, quickly washing it away. Thus a felled forest may provide a few years of crops, but soon becomes more or less lifeless. The native peoples of the world's remaining rainforests are only too aware of this fact. For generations they have felled small areas of forest, planted crops for a few years and then moved on elsewhere. The small-scale clearings created have quickly been taken back into the forest before lasting damage could be done. At the other extreme, clear felling for timber creates huge empty areas that the forest can never recolonize, areas that are doomed.

With clear felling the rainforest disappears and along with it go the birds and other wildlife that are dependent upon it. Individual birds will simply fly away as their nesting grounds are destroyed, but that does not mean that they will survive. Birds are highly territorial, and displaced individuals just cannot move in next door, as it were, since next door is already full. But just as populations are inevitably reduced by the destruction of their habitat, so too are individual species actually lost. Third World timber companies are not noted employers of ornithological consultants to ascertain the bird populations of an intended clearance area. So it is quite impossible to know what is being lost. Suddenly, perhaps, a species is noted by a competent ornithologist as being scarce, or even absent, from a particular area, and that may be the first we know of its demise.

There seems little doubt that the rainforests are doomed and that many species of birds will disappear as a result. A few isolated stands may remain as monuments to what was once a magnificent wilderness. Here, well-guarded and protected, species will survive in glorious isolation and bird-watchers will travel from far and wide to see the remnant populations. A tourist industry may arise from the ashes of the rainforest, but the rain-forest proper will have gone. Campaigns to save these fabulous regions of the world are definitely drawing public opinion to the importance and plight of these areas, they fulfil an important function. But campaigns will not, of themselves, save the forests. Only governments and international agencies – the European Community, United Nations, Group of Seven, and others – can actually achieve on-the-ground results.

The world's rainforests are located in what are generally called Third World or developing countries. At the beginning of the 1990s virtually every country endowed with such forests is deeply in debt to foreign governments and/or international banks. The countries of Latin America, for example, owe billions of dollars to the banks of Europe and North America. They are never likely to be able to repay these debts, and are hard put even to pay the annual interest charges. Already there is a public campaign in the developed "donor" countries to wipe out the debts and leave the Third World free to develop without such a heavy burden. But, supposing the governments and banks could be persuaded to swop what are, after all, bad debts for areas of rainforest, no matter how remote? The banks would benefit from an enormous public relations coup and be seen to be as "green" as they could ever wish to be, and at no further cost. Of course, someone would have to ensure that the areas purchased were properly surveyed and that illegal felling did not take place, but such costs would be within the scope of international conservation organizations and a likely field for cooperative sponsorship. Readers will, I trust, forgive this simple plan being floated in what is, after all, a book on birds.

If the rainforests are the most obviously threatened of the world's habitats they are far from being alone in this respect. Much of savannah Africa is being converted from productive grassland to arid semidesert. The process is already under way in the Sahel. Here, along the southern edge of the Sahara, the failure of rains and crops has gripped the public imagination

Apparently safe birds make up only 25 per cent of the world's birds. The rest are either declining or actually endangered.

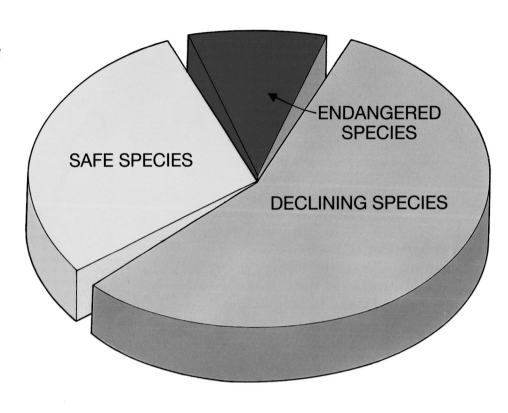

as never before. Films of starving children appear so regularly on our television screens that we are likely to become immune to suffering.

The problem is that the lack of rains means lack of grass and lack of crops, so that in order to survive the local people consume their next season's seed and overgraze what little vegetation remains. The Sahel has never been a productive place to live – it is at the very margins of human survival – but with the reduction of infant mortality the population has boomed. When an increasing population is coupled with lack of crop-producing rains, one has the elements of a disaster. Add in the odd civil war and the recipe is complete disaster.

The Sahara is extending southwards through the African savannahs at an alarming rate. Huge areas that were previously just about inhabitable are being turned to wastes of sand. Yet this same area was also a home to birds – and most importantly, to a huge number of small birds that spend their winters in Africa and their summers in northern Europe. Indeed, the first visible sign of the famine that was affecting the people of the Sahel was the "silent spring" in Europe when Whitethroats, Sand Martins and other birds that winter in the Sahel failed to return north to breed. To a huge number of Europe's avian summer visitors the Sahel is a refuelling area before or after crossing the Sahara. If these fattening grounds disappear then the birds are in trouble. So it is not only the local birds, but also those that come to winter or pass through that are seriously affected when an area of savannah turns to desert.

One can search for and find the causes that change an area in this way, and overgrazing and overpopulation come readily to mind. But knowing the cause does not create the cure. Life is always precarious in an area like the Sahel and it needs only a minor change to lead to disaster.

If the overall future of birds depends on the future activities of man and the global effect that man has on the planet, this does not mean that there will not be changes on a smaller scale. Some of these will be quite natural and species that are better equipped will replace those that are less able to deal with a particular set of circumstances. There will be natural changes in distribution as minor changes in climate affect birds at the margins of their ranges. Already we have seen the disappearance from England of the Wryneck and Red-backed Shrike, species for which southeastern England represented just about tolerable conditions. There is no hint in either case that anything other than natural causes were at work and both species remain well established in mainland Europe.

There are, however, some species and groups of birds that find it increasingly difficult to live in a human-dominated world. In these cases it is not the destruction of habitat but straightforward persecution that is to blame. The falconers of Arabia and the Middle East have enjoyed fantastic wealth during the present century, leading to an enormous increase in the demand for falcons and for prey at which to fly them. The effects have proved detrimental to both falcons and prey alike, but to none more so than the world population of the Houbara Bustard. In a few short years this medium-sized, desert-dwelling bird has been almost wiped out in the Middle East, and the falconers have turned their attentions to the populations of bustards in Pakistan and, latterly, in northwestern India. Only a political outcry managed to save these birds in the countries concerned.

Other bustards face other problems. How can a 40 pound plains bird manage to survive in a Europe so dedicated to food production? The Great Bustard breeds among the plains of Iberia, Austria, Hungary and at a few other sites in western Europe. It prefers natural grasslands wherever they exist, but will turn to cereals in the absence of grass. In grassland some nests are destroyed by domestic stock, but unfortunately nearly all nests in cereals are doomed due to modern agricultural machinery. One of the great European strongholds of the Great Bustard is the plains around Cáceres in Extremadura in central Spain. Even here, however, the amount of plowed land seems to increase year by year, leaving smaller and smaller parcels of grassland for the bustards. Only as I write has the plight of the Bustard at its major Spanish breeding grounds received the attention it deserves. The International Council for Bird Preservation (ICBP) has made 1992 the year of the Spanish Steppes and one can only hope that their efforts will persuade the Spanish government to take the project on board.

That the larger and more spectacular birds suffer most from direct human persecution is perhaps understandable. The falcons, bustards, cranes, egrets gamebirds and others have all variously been affected by hunting. Smaller birds are generally more affected by habitat destruction, though in the Mediterranean even the smallest of birds are still considered fair game.

In France and Italy virtually everything that flies is likely to be shot for the pot. In France, and elsewhere, huge numbers of birds are netted annually to provide such delicacies as larks' tongue pâté. In southern Spain, winter markets offer Song Thrushes by the dozen, with the occasional Blackbird mixed in by accident. In Cyprus the particularly obnoxious technique of liming is still widely used, despite well-meaning laws banning the practise. In southern Europe small birds are massacred in huge numbers to be pickled or served on toast as local delicacies. Even television cookery programs show such gastronomic delights without comment.

Although there is naturally widespread condemnation of such onslaughts on the populations of small birds, even Britain, the most conservation conscious of European nations, allows many species of waders to be shot as game. Frankly, I have never been able to appreciate why Snipe should be shot and Song Thrushes not, or why it is alright to shoot Woodcock but not to shoot Blackbirds. It does not follow that we should be against shooting as such. There are good reasons to allow, in Britain at least, the shooting of, say, Woodpigeons and Magpies.

Woodpigeons are a serious agricultural pest that no amount of shooting could possibly endanger, yet even here the "sportsmen" have got it wrong. The Woodpigeon is traditionally shot in the autumn, when numbers are highest. If we seriously wanted to reduce their numbers we would ban shooting from the end of the breeding season to the following spring and let nature help us in our task. Winter is the lean season for all birds. Therefore, the larger the number of Woodpigeons that feed in autumn, the smaller the stock of food that remains to see them through the winter. In this way natural mortality is higher in the late winter when food runs out and the fewer birds there are to breed the following summer. This number could then be further reduced by spring shooting.

Possibly the same argument applies to the Magpie, though no parallel study has been undertaken. Here the problem is indiscriminate nest robbing, for Magpies not only rake out the eggs and young of many small birds, but they also take eggs and chicks of gamebirds. At one time gamekeepers' gibbets were always full of Magpies. Today there are many fewer gamekeepers and considerably more Magpies. They have moved from the suburbs to city centers during the past 20 years and are more abundant now than ever before.

Birds are, then, widely persecuted virtually throughout the world. They are killed for sport, for food, for their feathers and, in Antarctica, penguins are boiled down en masse for their oil. This last is a particularly obnoxious practise, showing, as it does, the greed and shortsightedness of our human species.

The future of birds depends progressively on mankind – particularly on our attitude towards this planet and the natural habitats that we care to protect or destroy, and so on our attitude to birds themselves. Unless some action is taken to limit the human population, man or the planet – or both – face disaster. As a consequence, while the number of bird species will undoubtedly decline, those species that actually benefit from man's profligate treatment of nature will flourish. There will never be a silent spring, but there may be a spring punctuated only by the chirrups of House Sparrows and the wheezing of Starlings.

A colleague who read this chapter thus far stated that he considered going home and slashing his wrists. So, lest it be thought that all is gloom and despondency, let us take a look at the plus side where, perhaps, we may find some cause for optimism.

There is no doubt that the "green message" has been heard and that, in many parts of the world, people have become more aware of the health of our planet. By the end of the 1980s there were even "green" members of national parliaments in some western countries and it was *de rigeur* for any candidate for elected office to profess a caring attitude toward the environment. Holes in the ozone layer, acid rain, land and water pollution – all were being confronted and there was cause for optimism. Yet, at the same time, western democracies, or at least their governments, were becoming progressively more concerned with their own domestic economies, with inflation, unemployment and recession; with standards of public health care; with the dynamic foreign affairs situation following the Gulf War, nuclear proliferation, civil war in Yugoslavia and the dangers of the breakup of the former Soviet Union. Once again the environment was pushed into the background of public thought and debate. Perhaps public concern about such matters can only come to the fore when there is nothing else to worry about.

In 1983 I was able to tell the stories of the birds that had been able to return to breed in Britain after being exterminated during the nineteenth and early twentieth centuries. *Birds that Came Back* was in many ways a celebration of bird conservation showing how a more enlightened attitude towards our countryside had produced more favorable conditions for birds. The return of the Avocet, Bittern, Marsh Harrier, Black-tailed Godwit, Ruff and Osprey to breed and prosper in Britain is well known. As this is being written there is cause for more celebration as reintroduction programs, with White-tailed Eagle and Red Kite, begin to show signs of success. Most of these stories reflect admirably on the Royal Society for the Protection of Birds and their dedicated bands of enthusiasts, but also on naturalists in other countries, where changing laws and greater degrees of protection and public interest have provided the "reservoirs" of surplus birds to permit both colonization and reintroduction. In this connection one thinks of the remarkable changes in Norway where, from having a bounty on its head in the 1950s, the White-tailed Eagle has prospered to enjoy its major European stronghold. It was from this source that the birds reintroduced to Scotland derived.

Elsewhere other measures of active conservation have met similar success. The boom in numbers of the Whooping Crane in North America is totally due to the establishment of a captive breeding and release program that has, in a matter of ten or so years, trebled the population. At only 133 birds the situation is far from secure, the problem far from solved, but at least there is a sense of optimism in crane conservation circles.

Similarly, the drastic action of taking the last few wild California Condors into captivity in the 1980s, and the consequent political debate that ensued, would seem to be paying dividends. Captive birds have managed to breed successfully and there is already talk of an imminent release, though if they are to return to the polluted and dangerous environment whence they came then there can be little real hope of success. Yet these great birds were once found from Washington southward through Oregon to California and beyond. Surely there must be some ranges within this vast area where a successful reintroduction could be organized?

Perhaps the greatest of all bird conservation successes is one that followed the greatest of all bird disasters. In the 1950s and 1960s millions, perhaps billions, of birds were killed by the use of chlorinated hydrocarbon pesticides such as DDT. These new-to-science compounds, used to protect both seeds and crops, were remarkably toxic and persistent. So not only did they kill agricultural pests but, via infected insects, found their way "persistently" into the food chain. This mammoth blunder led Rachel Carson to produce her famous book *Silent Spring* and governments to take action. Voluntary bans on the use of the chemicals in Britain came early and, as a result, the extent of the slaughter was less than elsewhere, where legal bans had to be imposed. As a result Britain has become one of the major strongholds of the Peregrine Falcon anywhere in its cosmopolitan breeding range.

One of the troubles with chlorinated hydrocarbons is the length of time that they take to break down, their persistency. Thus infected insects or treated seeds might be eaten by a bird in less than lethal quantities. The bird, in turn, might be caught by a predator. Over a period of time pesticides accumulated in the bodies of the predators leading to death or serious malfunction. While all predators suffered by being at the end of the food chain – the end users – Peregrines, existing on a diet of seed-eating pigeons, were particularly badly hit. The population of these falcons in Britain was halved, while in North America the bird was brought to the verge of extinction.

Prompt action saved the British Peregrine population and 25 years later the birds have reached a level quite unknown this century. They have spread back to many haunts from which they had been absent since 1940, including the heavily developed south coast of England. In North America, where the situation was far more serious, a program of captive breeding has proved immensely productive and a reintroduction scheme has seen birds back at many of their classic sites.

Yet no sooner had these birds begun to recover than they faced yet another danger – the demands of falconers, especially those of the immensely rich oil sheikdoms of the Middle East. What had once been the sport of princes suddenly became the sport of the Arab entrepreneur as well. With money being no object the price of falcons, especially the more powerful Sakers and Peregrines, soared and crooks found it more profitable to steal fledglings than rob banks or mug old ladies. As a result the Peregrine eyries of Britain, in particular, have been subjected to a series of

assaults with young birds being illegally taken and smuggled to the Middle East via other parts of Europe. Of course, every year some of these criminals are caught and duly punished, but this does little to deter those out for a quick profit. Indeed, the only result is to put up the price of a young, wild-taken bird in the Gulf.

Despite this relatively new threat, the Peregrine population on both sides of the Atlantic continues to boom and is a cause of celebration in conservation circles. It is, however, like the stories of wild birds returning to breed in Britain, like the successes with Whooping Cranes and California Condors, no more than a drop of optimism in a sea of despondency. Everywhere through the world natural habitats and their associated bird populations are being continuously destroyed. Perhaps my colleague should slash his wrists after all, though I'd be inclined to enjoy a longer life spent at least attempting to save species for as long as humanly possible!

Just why the snipe, seen here at its grassland nest, should be regarded as fair game for hunters, while other similarly sized birds like the blackbird are not, causes considerable controversy between conservationists. Perhaps we should start a "Save-the-Snipe" campaign?

PICTURE
ACKNOWLEDGEMENTS

The author and publishers wish to thank the following sources for use of photographs:

Survival Anglia Library:
Peter Hawkey, pages 3, 41, 59, 211, front jacket; Jeff Simon, page 6/7; Ton Nyssen, page 9; Des & Jen Bartlett, pages 10, 16, 34 (top), 39 (bottom), 85, 104, 110, 111, 116, 118, 122, 156, 162, 194, 197 (top); Annie Price, pages 11 (top), 159 (top & bottom), 171 (bottom), 174; Jeff Foott, pages 11 (bottom), 70, 82, 83, 86, 89, 90, 91, 95, 99, 132, 204, 207, 208; John Harris, pages 15, 80, 178, 219; Dr F. Köster, pages 19 (top & bottom), 20, 176, 182, 184, 185 (bottom), 189 (bottom); Dieter & Mary Plage, pages 25, 79, 81, 137 (top & bottom), 145 (bottom), 148, 149, 199; Dennis Green, pages 29, 43, 52, 62; Claude Steelman, pages 33, 50, 94 (top); Mike Tracey, page 38; Matthews/Purdy, page 39 (top); Jozef Mihok, pages 47, 203; Vivek Sinha, pages 49, 131, 141 (bottom), 142, 145 (top), 189 (top), back jacket; Marek Borkowski, page 51; Michel Strobino, pages 53, 209; Joe B. Blossom, pages 54, 202; Rick Price, pages 58, 171 (top), 172; Terry Andrewartha, pages 60, 75; Andrew Anderson, page 65; John Lynch, page 68; Alan Root, pages 98, 164, 177, 183; Joan Root, pages 102, 103; Cindy Buxton, pages 105, 127 (bottom); Bruce Davidson, pages 109, 117, 124, 190 (bottom); Ian Wyllie, page 112; Deeble/Stone, page 125; Joanna Van Gruisen, page 135; Tan Ju Hock, pages 139, 191; Mick Price, page 153; Joel Bennett, page 175; Maurice Tibbles, pages 186 (top), 210; Frances Furlong, page 212.

Ardea London:
P. Morris, pages 12, 192; Ardea London, pages 13, 196; C. R. Knights, page 27; J. M. Labat, pages 28, 66; Peter Steyn, pages 34 (bottom), 44, 46, 100, 106, 115, 126; S. Roberts, pages 36, 94 (bottom); R. J. C Blewitt, page 37; Eric Dragesco, page 42; J. A. Bailey, pages 48, 61, 78; G. K. Brown, page 56; Richard Vaughan, page 57; Kenneth W. Fink, pages 67, 130, 134, 136, 185 (top), 186 (bottom), 187 (bottom); Donald Burgess, page 71; Martin W. Grosnick, page 72; C. & J. Knights, page 74; W. Stribling, page 76; B. L. Sage, page 77; Ian Beames, page 88; François Gohier, pages 92, 200; Wardene Weisser, pages 96, 144, 181; Alan Weaving, pages 107, 113; R. M. Bloomfield, page 108; Dennis Avon, pages 114, 119, 120, 128, 143; John Wightman, page 121; John Daniels, page 123; Anthony & Elizabeth Bomford, pages 127 (top), 129; Joanna Van Gruisen, pages 138, 150, 151, 152; John Gooders, page 141 (top); Don Hadden, pages 147, 154, 157; Jean-Paul Ferrer, pages 155, 163, 166 (bottom), 167; Eric Lindgren, page 158; Hans & Judy Beste, pages 160, 165, 168, 169; M. D. England, pages 187 (top), 193, 201; John Mason, page 190 (top); A. Greensmith, page 198; Clem Haagner, page 197 (bottom).

Humboldt University, Natural History Museum, Berlin: page 14.

INDEX